A review and methods for the use of epilithic diatoms for detecting and monitoring changes in river water quality 1993

F E Round

Methods for the Examination of Waters and Associated Materials

London: HMSO

First Published 1993

ISBN 0 11 752771 8

Within the Methods for the Examination of Waters and Associated Materials series are four-ring binders suitable for use in storing reports. These are available from HMSO Price £4 (ISBN 0 11 7514373).

HMSO

Contents

About this booklet 4

About this series 5

Warning to users 6

1 Introduction 7
- 1.1 Reasons for using diatoms for monitoring changes in river water quality 7
- 1.2 The diatom cell 8
- 1.3 The habitat 8

2 Diatoms and water quality 10
- 2.1 Previous relevant publications and reviews 11
- 2.2 Physical features 11
- 2.3 The problem of silt and macroscopic growths on stones 12
- 2.4 Sources of taxonomic information 14
- 2.5 Sources of ecological information in floras 14
- 2.6 Introduced substrata 15

3 The method 17
- 3.1 Method using epilithic diatoms for monitoring changes in river quality 17
- 3.2 Additional comments 19
  - 3.2.1 Sampling—Field 19
    - 3.2.1.1 Selection of site 19
    - 3.2.1.2 Selection of material 20
  - 3.2.2 Sampling—Laboratory 20
  - 3.2.3 Cell counting 21
  - 3.2.4 Minimum numbers 21
  - 3.2.5 Counting errors 22
  - 3.2.6 Tabulation of counts 22

4 Indices 22
- 4.1 Kothe's index 23
- 4.2 Cairns' sequential comparison index 23
- 4.3 Descy's index 23
- 4.4 Coste's index 24
- 4.5 Leclercq & Manquet's index 28
- 4.6 Lange-Bertalot's index 29
- 4.7 Watanabe's index 29
- 4.8 Palmer's index 30
- 4.9 Generic index 30
- 4.10 Data from British rivers 30
- 4.11 Suggested sub-division of river zones based on data from British rivers and related to work in other regions 33
- 4.12 General discussion of indices 37

5 Future work 39

6 Appendices: 41
- 1 Example of recording sheet 41
- 2 Light micrographs of species 42
- 3 Scanning electron micrographs 44
- 4 Notes on species 47
- 5 Light micrographs of communities 51
- 6 Saprobic and indicator values 53
- 7 Defined area sampling 54

7 Glossary 55

8 References 57

Address for correspondence 64

Members of the committee assisting with this booklet 65

About this booklet

It has often been noted that whilst chemical analyses are invaluable as evidence for a prosecution for pollution, biological monitoring is also needed. Biological monitoring can provide proof of damage; but more importantly it can also advise as to when to take the vital chemical sample, and it is gradually becoming apparent that one day, given further study, it may be able to indicate the type of pollutant. This would be a great help when there could be a large number of possible types of pollutant from different sources.

Diatoms are abundant throughout rivers and have been used in many countries as indicators of water quality. Usually the work has been associated with measures of organic and/or nutrient levels, but it is possible to relate diatom species to other environmental factors, eg heavy metal contamination. This account contains a brief discussion of diatoms in rivers and discusses in greater detail the use of one diatom community—that growing abundantly on stone surfaces—for detecting and monitoring changes in water quality.

The actual method is found in chapter 3.1, with additional information in 3.2 and in the appendices. Interpretation is discussed in chapter 4. However, as the method is unfamiliar to most aquatic scientists, newcomers are advised to read the chapters in the order given, so that they have enough background before starting to use the method. Although tested and proved in at least four laboratories, it is still regarded as tentative, firstly because it is still being developed, and secondly, because there is still much work to be done relating the occurrences of diatoms to individual environmental factors.

'Society's interest in pollution is centered on its effects on living organisms, and the assessment of pollution is principally a biological problem'.

Cairns & Dickson (1971),
Journ. Wat. Pollut. Control Fed. **43**

'merely knowing the concentration of the chemical (or other potential pollutant) is not likely to produce useful management information.'

Cairns & van der Schalie (1980),
Wat. Res. **14**

'micro-organisms are often the most direct indicators of environmental change.'

Dussart (1982),
Environ. Poll. **27**

'Given a river continuum with longitudinal differences in physical and chemical parameters (current, light, pH, nutrients) qualitative differences in the distribution of algal populations should be expected.'

Keithan, Lowe & DeYoe (1988),
J. Phycol. **24**

About this series

This booklet is part of a series intended to provide recommended methods for determining the quality of water and associated materials. In addition, short reviews of the more important analytical techniques of interest to the water and sewage industries are included.

In the past, the Department of the Environment and its predecessors, in collaboration with various learned societies, have issued volumes of methods for the analysis of water and sewage culminating in 'Analysis of Raw, Potable and Waste Waters'. These volumes inevitably took some years to prepare, so that they were often partially out of date before they appeared in print. The present series is published as a series of booklets on single or related topics, thus allowing for the replacement or addition of methods as quickly as practicable without the need for waiting for the next edition. The rate of publication is also related to the urgency of the requirement for that particular method.

Although ideally, all methods published should be fully tested, this is not often possible without delay in publication. Furthermore, the limit of detection, range, precision and interference effects applying to instrumental methods can depend on the actual instrument used, as well as on sample type, reagent purity and operator skill, etc. Even methods tested in many laboratories have been known to acquire problems, for example, when new products appear (introducing new substances into effluents), when changes in production methods affect reagent quality, or when the method is used to analyse new types of sample (despite apparent similarity to samples already evaluated). As a guide, the following categories have been given to methods:

(i) tested, usually in five or more laboratories
—no grade indicated;

(ii) tested in one to three or four laboratories
—Tentative;

(iii) evaluated, but not fully tested, but publication is urgently required
—Note;

(iv) tested and found to be satisfactory by several laboratories, but in the opinion of experts requires a high degree of skill or has some other difficulty such that the method would be replaced if a better method were discovered.
—Provisional.

The aim is to provide as complete and up to date a collection of methods and reviews as is practicable, which will, as far as possible, take into account the analytical facilities available in different parts of the United Kingdom, and the quality criteria of interest to those responsible for the various aspects of the water cycle. Because both needs and equipment vary widely, where necessary, a selection of methods may be recommended for a single determinand. It will be the responsibility of the users and senior technical staff, to decide which method to use for the determination in hand. Whilst the attention of users is drawn to any special known hazards which may occur with the use of any particular method, responsibility for proper supervision and the provision of safe working conditions must remain with the user.

The preparation of this series and its continuous revision is the responsibility of the Standing Committee of Analysts (to review Standard Methods for Quality Control of the Water cycle). The Standing Committee of Analysts is a committee of the Department of the Environment set up in 1972. Currently it has nine working groups, each responsible for one section or aspect of water cycle quality analysis. They are:

1.0 General principles of sampling and accuracy of results
2.0 Microbiological methods
3.0 Empirical and physical methods
4.0 Metals and metalloids
5.0 General nonmetallic substances
6.0 Organic impurities
7.0 Biological monitoring
8.0 Sewage works control methods
9.0 Radiochemical methods.

The actual methods and reviews are produced by smaller panels of experts in the appropriate field, in cooperation with the working group and the main committee. The names of those associated with these methods are listed at the back of this booklet.

Publication of new or revised methods will be notified to the technical press. A current list of publications can be obtained from the Secretary.

Every effort is made to prevent errors from occurring in the published text. Correction notes and minor additions to published booklets not warranting a new booklet in this series will be issued periodically. However, should any errors be found, please notify the Secretary.

Dr D WESTWOOD
Secretary

22 January 1992

Warning to users

The analytical procedures given in this booklet should only be carried out by competent trained persons, with adequate supervision when necessary.

Local Safety and COSHH Regulations must be observed.

Laboratory procedures should be carried out only in properly equipped laboratories.

Field operations should be conducted with due regard to possible local hazards, and portable safety equipment should be carried.

Care should be taken against creating hazards for one's self, one's colleagues, those outside the laboratory or work place, or subsequently for maintenance or waste disposal workers. Where the Committee have considered that a special unusual hazard exists, attention has been drawn to this in the text, so that additional care might be taken beyond that which should be exercised at all times when carrying out analytical procedures. Reagents of adequate purity must be used along with properly maintained apparatus and equipment of correct specifications. Specifications for reagents, apparatus and equipment are given in manufacturers' catalogues and various published standards. If contamination is suspected, reagent purity should be checked before use.

The best safeguard is a thorough consideration of hazards and the consequent safety precautions and remedies well in advance. Without intending to give a complete checklist, points that experience has shown to be often forgotten include: laboratory tidiness, stray radiation leaks (including ultra violet), use of correct protective clothing and goggles, removal of toxic fumes and wastes, containment in the event of breakage, access to taps, escape routes, and the accessibility of the correct and properly maintained first-aid, fire-fighting and rescue equipment. Hazardous reagents and solutions should always be stored in plain sight and below face level. Attention should also be given to potential vapour and fire risks. If in doubt, it is safer to assume that the hazard may exist and take reasonable precautions, rather than to assume that no hazard exists until proved otherwise.

There are numerous handbooks on first aid and laboratory safety. Among such publications are: 'Safe Practices in Chemical Laboratories' and 'Hazards in the Chemical Laboratory', issued by the Royal Society of Chemistry, London: 'Safety in Biological Laboratories' (Editors Hartree and Booth), Biochemical Society Special Publication No 5, The Biochemical Society, London, which includes biological hazards; and 'The Prevention of Laboratory Acquired Infection', Public Health Laboratory Services Monograph 6, HMSO, London.

It cannot be too strongly emphasised that prompt first aid, decontamination, or administration of the correct antidote can save life; but that incorrect treatment can make matters worse. It is suggested that both supervisors and operators be familiar with emergency procedures before starting even a slightly hazardous operation, and that doctors consulted after any accident involving chemical contamination, ingestion, or inhalation, be made familiar with the chemical nature of the injury, as some chemical injuries require specialist treatment not normally encountered by most doctors. Similar warning should be given if a biological or radiochemical injury is suspected. Some very unusual parasites, viruses and other micro-organisms are occasionally encountered in samples and when sampling in the field. In the latter case, all equipment including footwear should be disinfected by appropriate methods if contamination is suspected. If an ambulance is called or a hospital notified of an incoming patient, give information on the type of injury, especially if poisoning is suspected, as the patient may be taken directly to a specialized hospital.

Safety while Sampling

Prior consideration must be given, especially when sampling in confined spaces or where access is difficult, to guard against suffocation, drowning, falls and poisoning or infection by ingestion, inhalation or skin contact.

Good Laboratory Practice

The Department of Health issues a booklet entitled: Good Laboratory Practice; the United Kingdom Compliance Programme, 1989. This can be obtained by writing to that Department in London. It deals chiefly with toxicity studies, but much can be applied to analytical chemistry.

1 Introduction

Diatoms are the most abundant autotrophic organisms in rivers, colonising all available surfaces (loosely termed the benthos) and in slow-flowing rivers also colonising the water column (plankton). Total diatom biomass, their fixation of carbon, release of oxygen, loss by flow into the food chain or senescence have been investigated by relatively few workers, though these features are at the centre of biological activity in rivers. For example, Gale *et al* (1979) estimated that in samples from stones in deep water in the Susquehanna river the diatoms comprised 95% of the biomass dropping to 65% in the more polluted zones. The presence of diatoms is frequently obvious to the naked eye as a brown coating on all underwater surfaces which, when touched, proves to be coupled with a mucilage exudate; this mucilage also has important properties affecting the river ecosystem. The diatoms are a major source of food for the protozoa, invertebrates and young fish. The diatoms and their extruded mucilage also act as an ion exchange system (mainly in low pH waters) influencing the chemical composition of the water.

1.1 Reasons for using diatoms for monitoring changes in river water quality

The use of diatoms in water quality assessments has been described in many European studies and much support has been given to this approach. Leclercq (1988) tested three systems in Belgium, a chemical, a macro-invertebrate and a diatom index, and concluded that the diatom system gave the best results. The advantages of indices based on diatoms are usually quoted as:

(a) universal occurrence throughout all rivers;
(b) rapid, easy field sampling;
(c) diatom species are very sensitive to water quality (chemistry), eutrophication and pollution;*
(d) the cell cycle is rapid and they react quickly to perturbation of the environment (Oemke & Burton (1986) report diatoms colonising glass slides within 2h of immersion);
(e) diatoms are relatively insensitive to physical features of the environment;
(f) cell counting by microscopic techniques is rapid and accurate;
(g) cell numbers per unit area of substratum are enormous, making random counts excellent assessments of the flora. Dussart (1982) gives the sensible advice to sample stones with the thickest algal growth (brown not green pigmented growth):
(h) the ecological requirements of diatoms are in many cases better known than any other group of river organisms;
(i) the number of species needed for assessment need not exceed 50 (Lange-Bertalot 1979). Further evidence for this kind of number is provided by Engelberg (1987); he found that 43 species made up 85% of the diatoms on stones (and plastic). The vast majority of papers report 1–5 species dominant at any one site;
(j) unlike invertebrates, they do not have specific food requirements, specialised habitat niches, nor are they governed to a major extent Statzer & Higler (1986) by stream flow.

These advantages can be supplemented by several others eg

(1) the slides made for identification of diatoms can be made permanent and can be retained so that comparisons can be made over time and referred to by other workers.
(2) the cleaned material can be preserved, re-checked, and distributed to other laboratories, etc;
(3) diatoms are easily cultured and are therefore amenable to experimental verification of their reaction (tolerance) to nutrients (see also experiments by Fairchild & Lowe (1984), Fairchild *et al* (1985), Hürlimann & Straube (1991) on *in situ* chemical manipulation, pollutants etc); great care is needed when experimentation is undertaken since diatoms in the natural habitat are affected at lower concentrations of chemicals (eg NH_4–N), than that used in experimental channels;

* There have been a few experimental studies in which rocks were transplanted from clean to polluted sites and *vice versa* eg Iserentant & Blanke (1986) showed that the diatoms on rocks from clean water were very susceptible to pollution but the transfer of polluted rocks to clean water did not immediately eliminate the flora—the pollution indicators were still present 45 days later—though it is not clear from the data whether or not the cells recorded were live.

(4) the same diatom species are found in many parts of the world and hence a massive data base is already available;
(5) diatom populations recover rapidly after depopulation by flooding, Fisher *et al* (1982), and probably after pollution events;
(6) only dominant species need to be used;
(7) the identifications and counts can be done by non-specialists, if they are provided with illustrated guides;
(8) type slides for checking microscopic details could be provided thus aiding standardisation between laboratories.

The aim of the present review, concentrating on the epilithon, is briefly to review available sampling methods and indices and to suggest accurate but simpler methods which can be used in the biological surveillance of rivers and yield additional indicators of water quality to supplement data from chemical, zoological and bacteriological surveys.

Why the epilithon? Simply because to date almost all methods based on diatom indices in use throughout the world concentrate on this community; though no-one has ever demonstrated by comparative studies that this diatom community is better than the others. Why select the diatom component of the epilithon? Simply because they are usually the dominant autotrophic group, the identification of diatom species is simpler than for other components (Cyanobacteria, Chlorophyceae), the diatoms are present throughout all rivers, and present throughout the year, they have no cell cycle complications, they are ecologically better known than any other group and permanent records (as microscope slides) can be made from every sample (if necessary). There have, in addition, been many studies using the diatoms cultivated on artificial hard surfaces placed into rivers. These methods will be dealt with briefly since some workers regard them as suitable alternatives to the natural stone substrata; they do provide valuable distributional data for the literature, and the method can be applied to certain specific problems.

1.2 The diatom cell

Diatoms are unicellular and unique amongst photosynthetic organisms in having a rigid multi-partite siliceous cell wall (often referred to as the frustule) around the cytoplasm. A diagrammatic cell and an expanded example together with the terminology describing the cell is given in Fig. A–C. Due to the method of cell division, diatom cells (frustules) get steadily smaller over a period of time and recognition of large and small cells of the same species is vital. Small cells finally undergo sexual reproduction first forming an auxospore (Fig 2a) and then usually a malformed cell (Fig 2b) which is only rarely seen in populations. At the next division, normal cells (Fig 2c) are formed which begin their cycle of reduction again (eg Fig 2d). For further details, see Fritsch (1948), or the most recent account by Round *et al* (1990) in which cellular structure is discussed in detail, and approximately 250 genera are described and figured.

1.3 The habitat

Distinct associations of diatoms occur closely bound to the micro-habitats, eg on sediment, sand, gravel/stone/bedrock or the macrophytic plants—the communities on these being designated as epipelon, epipsammon, epilithon and epiphyton*. Diatoms are usually the dominants in these surface habitats, but they are always accompanied to varying degrees by other algae—the total species running into several hundred. In addition each micro-habitat has bacterial, fungal and animal (Protozoa, nematode, oligochaete, insect larvae, etc) components, see Lock *et al* (1984). The composition of the diatom associations is influenced by water flow, river chemistry, eutrophication, pollution and grazing. The combination of these results in the potential for a large number of diatom associations, though for practical purposes these can be reduced to a manageable number.

* The collective term 'periphyton' is often used, but it is an imprecise term which should be avoided.

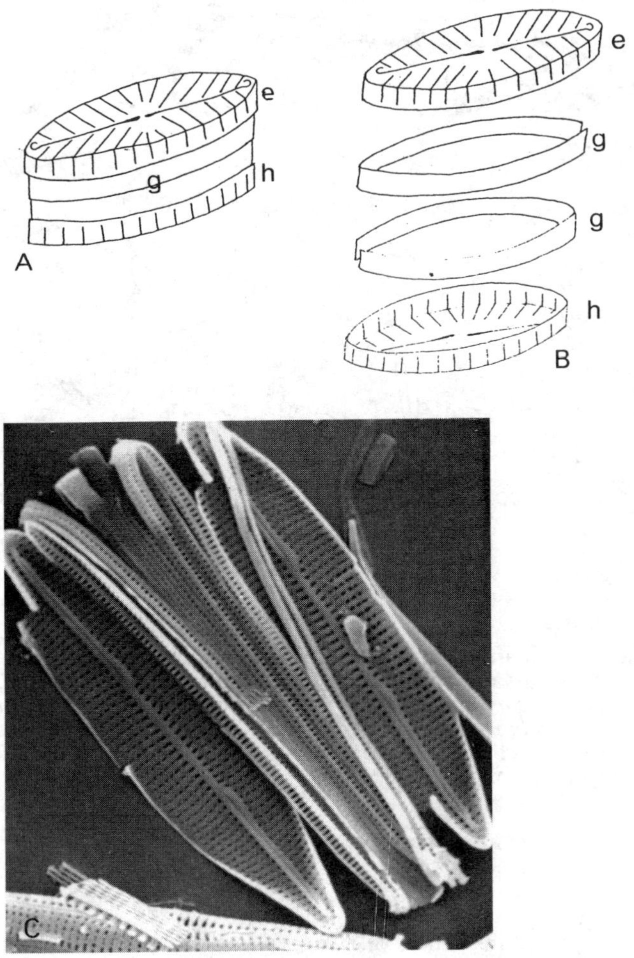

FIGURE 1

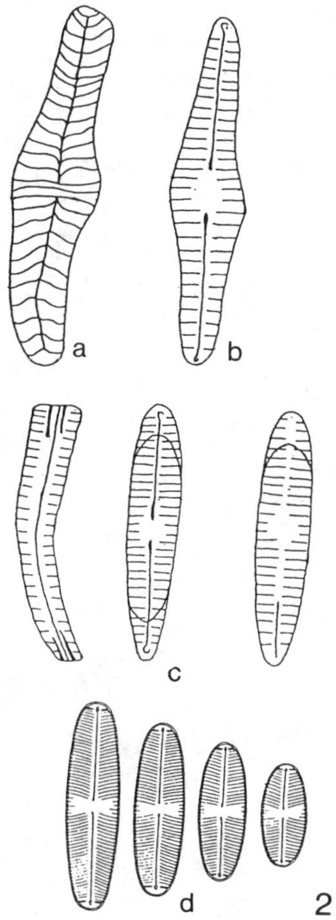

FIGURE 2

Fig 1. A diagrammatic representation of a diatom cell, (A) Epivalve (e), Hypovalve (h) and Girdle Bands (Copulae) (g). (B) The siliceous parts of the cell expanded. (C) Scanning electron micrograph of the diatom cell with the two valves (seen from the inside) and the girdle bands between, photographed after cleaning to remove the organic contents.

Fig 2. An auxospore (a) enclosed by siliceous bands. A somewhat distorted primary valve (b) arising from the germination of the auxospore. (c) a girdle view of the diatom *Rhoicosphenia* and the two valves with dissimilar markings. (d) an example of size reduction in a diatom during growth of a population.

The epilithon is the basis of the method in this handbook, but it is essential to bear in mind the other communities, since cells from these may contribute to or contaminate the epilithon. Contamination (usually by dead cells) is inevitable in a flowing water system and diatoms with their rather resistant siliceous cell walls may remain in the communities longer than other organisms. This fact has probably led to some confusion, eg Stevenson & Hashim (1989) to suggest that all the micro-habitats have a relatively similar flora. Techniques need to be used which minimise contamination and assess the structure of the discrete communities.

The epilithon, Fig 3, colonises stone surfaces—ranging from stable bedrock, through boulders, cobbles, pebbles, gravels, sand, silt (Wentworth scale). There are differences between the floras of the first four surfaces but they are slight. The smallest inorganic particles colonised are sand and silt and these support a rather different flora (the epipsammon and the epipelon) and will not be dealt with here. Few rivers have a uniform set of stones from the upper to the lower reaches but usually it is possible to find suitable substrata at most sites, since areas of riffles occur down most rivers and provide a supply of stones. The upper region can be more difficult to sample with only bedrock available and the lower reaches tend to be silted. In visits to over 1,000 sites in Britain, the author has rarely failed to obtain reasonable samples, Round (1989). There is slight evidence that stone size affects the biomass of diatoms—lower on small stones Duffer & Dorris (1966); the effect of stone size on floral composition still needs checking.

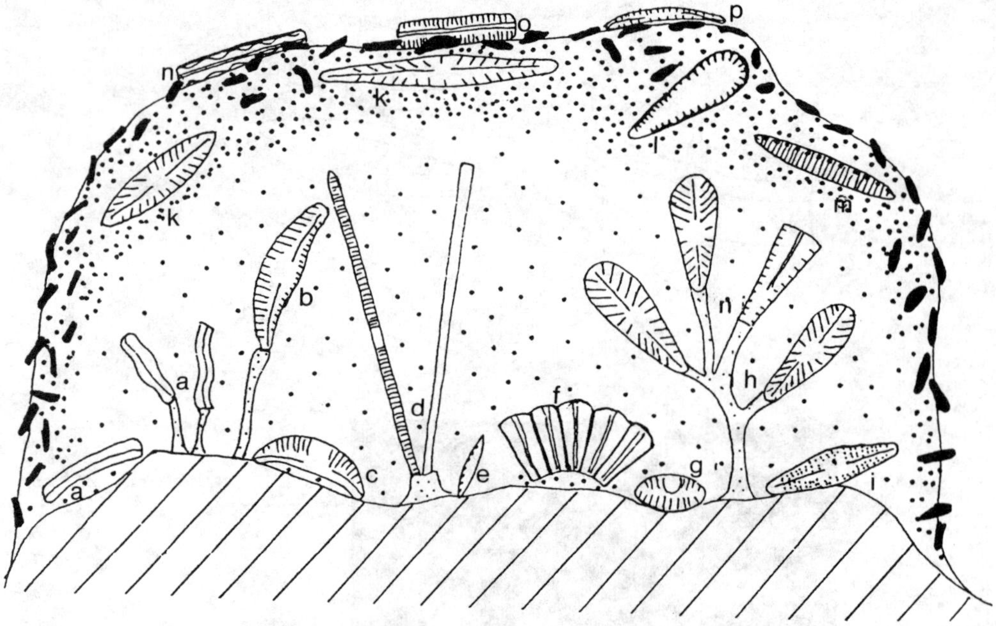

Fig 3. Diagrammatic representation of the epilithic diatom flora on a stone surface together with the overlying silt and its associated flora. a, *Achnanthes minutissima*, b, *Cymbella* sp, c, *Amphora pediculus*, d, *Synedra ulna*, e, *Nitschia* sp, f, *Fragilaria* sp, g, *Achnanthes lanceolata*, h, *Gomphonema*, i, *Navicula goepertiana*, k, *Navicula* sp, l, *Surirella* sp, m, *Nitchia* sp, n, *Cymatopleura* sp, o, *Navicula* sp, p, *Amphora* sp. This structured community has been compared to the layers of a forest (see 2.3).

2 Diatoms and water quality

Diatoms have been included in studies of river water quality since the early Kolkwitz & Marsson (1908) studies. Over the years, more and more species have been identified and used in systems designed for determining water quality based on biological data. Nowhere in the fairly extensive literature is there a thorough investigation of the various communities and most systems have not been sampled sufficiently precisely. The result is that many of the indices of water quality in the literature are not comparable, not merely because of contamination, but also because data from several communities has been combined—in some instances samples have simply been of 'river diatoms', eg Hartmann (1983). The epilithon is at least a community which can be sampled from a single spatial niche unlike, for example, the invertebrates. Even so, it has its complexities (see section 3). The epilithic diatom community has the great advantage that similar species occur in streams and rivers throughout the world and indicate roughly comparable environmental conditions.

The methodology for sampling and examination of diatoms will depend to some extent on the purposes of the survey. Detailed quantitative studies will require time-consuming, multiple samplings and statistical analysis of data. Overall, river surveys need simpler semi-quantitative techniques or even data based on indicator species, eg Wegl (1983). It is important to define clearly the object of the study and adapt the methods appropriately. There is no point in an academic approach which adds little useful information to that obtained by simpler methods. A few 'indicator species' dominate the epilithic flora and precision aimed at the less common species is pointless and time-consuming.

Some reports use identification to genus only, eg Gumtow (1955) and work with macro-invertebrates has shown that whilst species determinations give more information the less time-consuming family identification is often adequate: see Furse *et al* (1984) for an excellent discussion and many of the statistical techniques; for the latter see also Hellawell (1972) Kaesler *et al* (1974) also showed that reducing the data to two groups (Ephemeroptera and Coleoptera) gave as much information as using the total insect fauna. Care would have to be taken if identification of diatoms were taken only to genus or family.

2.1 Previous relevant publications and reviews

Of historical interest only as the first use of diatoms (along with other organisms) is the work of Kolkwitz & Marsson (1908). Later Butcher (1932, 1940, 1946, 1947) studied the epilithon (mainly) of British rivers and interestingly these studies stand alone as some of the most comprehensive undertaken. His work revealed that 'diatoms comprise the largest and most prevalent single group of algae in a river'; the floras differed between calcareous and non-calcareous rivers; communities in eutrophic and oligotrophic zones could be distinguished; adnate (attached), pedunculate (stalked) and free-living species were involved; there were seasonal growth patterns, but the dominant algae were always present and finally, that grazing was important. The only subsequent British work of any extent was that of Marker (1976) and colleagues at the FBA River Laboratory, but they were more concerned with biomass and physiological aspects and do not touch greatly on aspects of water quality; they tended to use chlorophyll values and consider that in the use of floristics it is more difficult to attain accuracy Casey *et al* (1981). Chlorophyll estimations (or other measures of biomass) or productivity measurements cannot be used easily in water quality studies. The Margalef index of pigment diversity, eg Cattaneo *et al* (1975) has rarely been used and considered unlikely to be of great value. There is nowhere a full account of the diatom flora in any British benthic river habitat; a comprehensive account of the variations in diatom flora along the length of a river; any relationship of flora to chemical conditions, nor to any form of eutrophication/pollution. This is in striking contrast to work elsewhere, especially on the Continent, in Japan and N. America where techniques have been evolved and data are available in many papers.

Reviews dealing in general with diatoms in rivers are Blum (1956), Fjerdingstad (1964, 1965), Patrick and Roberts (1979), Salden (1978), Sladeckova (1962), Sládecek (1973), Symoens (1957), Youngman (1971), Thorpe & Williams (1980), Round (1991) and Whitton *et al* (1991). Reynolds (1988) discusses the plankton in a stimulating article with valuable comments on the physical factors which may also impinge on the benthos.

The early work used the presence of diatoms or diatom associations as indicators of various water characteristics, ie they were used as indicator organisms. This system has not changed, but starting with Pantle & Buck (1952) the data were increasingly converted into numerical indices and these were related to organic pollution (saprobity). The indices have been extended, more species added, the mathematical/statistical bases argued and development is continuing eg in France and Belgium, Descy (1976, 1979, 1983), Descy, Empain & Lambinan (1982), Descy & Coste (1987, 1988, 1990), Leclercq & Manquet (1987) and in Japan, Watanabe (1986). The sampling technique is usually based on that designed by Douglas (1958) for mountain streams; but there are problems when this is applied to slower-flowing rivers and some workers report more general problems. Some of these studies involved around 400 species and require expert diatomists to handle the identification.

There is a serious dichotomy in almost all the work. The detailed work necessary on the flora (identification, counting, devising of indices, application of various statistical techniques, etc), is essential basic research, but much of it hinders the establishment of easily usable, rapid methods for detecting or monitoring changes in water quality. Concentration on numerical scales obscures much valuable data; indices can always be computed if only numbers are required, the reverse is not possible!

Workers are divided between those who sample *in situ* using devices based on Douglas's (1958) simple apparatus, eg Descy (1979, 1980, 1983, 1984) and those who collect stones and complete the sampling in the laboratory, eg Watanabe (1986, 1988). The latter technique has the advantage that observations on the stones can be made in greater detail in the laboratory than in the field. A further group of workers prefer to use artificial substrata.

2.2 Physical features

Water depth

On the whole, there is a reasonably uniform distribution of the diatom flora at any one sampling point, though slight differences can be found between stones from the shallows and those somewhat deeper—another aspect which needs further study. There are surprisingly very few studies on the effect of depth (light climate) on occurrence and growth characteristics of attached diatoms (in contrast to studies on planktonic species). Most river epilithic samples are taken at less than 1 m depth, but even if the maximum biomass occurs below 1 m it may not be necessary to sample at such depths if the indicator species are present in the shallow water—data in Round (1989) suggest they are. Many small streams are shallow so it makes sense to sample in shallow water thus maintaining comparability of light climate between sites. However, in some waters with high loading of solids, the light may be severely curtailed and in addition the stone surfaces may be covered by deposits. Stone surfaces will vary in length of time they have been exposed to colonisation—again, further study is needed. Such problems can be avoided if a selection of stones with visible brown coatings is obtained at each sampling site and the material bulked. There is no doubt that this obscures subtle variations but if what is needed is the average epilithic flora at the site, then this is obtained, see also Dussart (1982). Discretion is essential in river sampling due to the variability of reaches (in contrast to sampling standing waters), eg the effects of scour near the river edge or exposure during low flow. Common sense and direct observation of the stone surfaces usually leads to adequate sampling.

Place the stones in labelled polythene boxes without adding water. For general surveys there is no loss of data if stones touch each other, but for extremely accurate cell counting, studies of variation in the flora on different faces of stones, on different stone composition, etc greater care must be taken and individual stones carefully transported.

Current speed

This has some effect but it is probably only very obvious in the rhithron (the upper fast-flowing stretch) where scouring allows only the firmly attached forms to occur; *Achnanthes minutissima/A. microcephala* and *Diatoma hiemale* are most often quoted as the dominants here—they are also common in springs. Over much of the river system, enhanced flow will stimulate the flora and may also remove any contaminant silt/mucilage flora (see section 2.3). Several studies, eg McIntyre (1968), Reisen & Spencer (1970), Reiter (1986), Steinman & McIntyre (1987) have shown that colonisation (adhesion) is slower in fast flow than slow flow, but after a few weeks the full floral characteristic of the water quality builds up. Stevenson (1983) reports immigration rates onto tiles of 30–1,500 cells $cm^{-2}d^{-1}$; the rate was greater by a factor of 8 in more sheltered waters, Stevenson (1984a); it was also increased by a factor of 2 if the tiles were coated with agar, which would presumably have a similar effect to the bacterial slime film which forms initially on surfaces immersed in water.

Substrata

The geological origin of the stones may result in differing substratum composition down a river. This will involve many inter-related features of the stones, eg gross morphology, surface characteristics and chemical composition between slates/granites/sandstones/limestones, etc. There are a few studies on the effects of differing geology, Blinn *et al* (1980), Antoine & Benson-Evans (1985) and they are somewhat contradictory but the consensus seems to be that influence of stone type on the colonising diatoms is not extreme; both the above groups found sandstone the most favourable, especially in the early stages of colonisation. This is undoubtedly due to surface pitting since other workers have shown by scanning electron microscopy that colonisation occurs faster in the surface pits. Here again considerably more data are required whilst the differences are likely to be subtle and probably of little sigificance in broad water quality surveys—the chemical composition of the water over-riding the other effects. This contrasts with some statements in the literature although when the presence/absence of species is the crux of the methodology, it is the chemicals which seem to be the major determinands.

2.3 The problem of silt and macroscopic growths on stones

In all the slow-flowing reaches, silt tends to build up, in and on the outer layer of mucilage on the stones. This silt layer supports a diatom flora, some species of which may be characteristic of this microhabitat, but many are merely derived from the nearby epipelon (ie the community rich in diatom species which colonises the river sediment).

The species living in/on the silt are included in the epilithon by almost all workers, eg Descy (1983). This multiplies the number of species enormously and is misleading since when comparisons are made between silted and unsilted sites, the result may give the false impression that water quality differed at the two sites. What is the solution to the problem since it is not always possible to select stones which are unsilted? One solution is to wash the silt off the stones without removing too much of the underlying mucilage; there will, however, always be some contamination of the underlying epilithic flora. The only other solution is to examine the fresh material before removing the epilithon and note the dominant species of the silt flora; then discount these when making counts of the epilithon. On the other hand, it may be possible to utilise the silt flora as an extra source of information provided that it is present at all sites being monitored. The whole problem of the micro-strucure of the diatom communities on silted stones requires further study. It is possible to separate the flora into three components (from the silt, the inner mucilage and from the actual stone surface) and then clean these up (see section 3) and identify and count each separately (see Fig 3 for a diagrammatic representation of these layers).

Macroscopic growths of algae, bryophytes and angiosperms on stones are another source of contamination by diatoms. The first rule is to avoid sampling stones on which macroscopic growths are visible to the naked eye. But there are often immature stages on the stone surfaces. These may need to be removed in the laboratory by picking off with forceps. Again, as with the silt flora, complete removal may be impossible but so long as a rich epilithic flora is sampled, the epiphytic contaminants will be so diluted out that they will hardly figure in the counts. It may also be desirable to become familiar with at least the dominant epiphytic diatoms when a limited number of sites are being studied in detail.

The structure of this almost universal 'forest' of organisms, (Patrick & Roberts (1979) also used the term 'microphytic forest-like growths') in a mucilage matrix has been ignored by most workers and only Lock *et al* (1984) have attempted an investigation of it in rivers, though Hoagland *et al* (1982) illustrate it on artificial substrata which has been placed in reservoirs. The diatoms must be removed either by total removal (resulting in mixture of the 'forest' layers) or by removal layer by layer. The total removal technique is the one employed by at least the majority of previous workers. It is simple and crude (scraping, rubbing, brushing) followed by washing the loosened material off the stone surface into containers (centrifuge tubes are ideal). The area sampled can be delineated, eg Dussart (1982) used a sheet of cellophane with a 100 mm^2 hole in it—the area inside the hole was scraped and the material pipetted out; another technique was to dry the stone, mark out an area and coat around this with nail polish, allow to dry and then brush the uncoated area, Economou-Amili (1980). However there is little doubt that data in the literature have been obtained by workers who have removed varying amounts of the 'forest' layers. It should be noted that disturbance can remove the upper storey, Hudson & Bourget (1983) and of course the silt layer and use of diatoms from these layers (especially the silt flora) could affect indices which include these in the calculations; thus it may be better to rely only on the lower more firmly attached flora. The exact area brushed is less important than total removal of each layer—the basal layer of cells can be extremely difficult to remove and brushing often does not achieve this—rubbing with the forefinger or scraping with a scalpel may be necessary. The difficulty of removal is one reason why Marker (1976) turned to chlorophyll extraction techniques to estimate epilithic biomass but if floristic data are necessary and it is essential for determining indices or identifying 'indicator species', then adequate removal of cells from the stone surfaces is essential. An ultrasonic technique for removal of cells from stone surfaces was reported by Gale (1975), but it does not appear to have been followed up.

Previous workers have not distinguished the layers of diatoms in the 'forests'. Proceeding down a river from near its sources, the most likely scenario for this forest is (a) upper fast-flowing reaches where the basal, adnate flora (ground flora) is well-developed and may be the only component, (b) the slower flowing reaches which will support to varying degree both the adnate and the upright (usually stalked, ie pedunculate) species, (c) the slowest flowing and/or silted regions will in addition have a canopy of adventitious species growing in the upper layers of mucilage. **NB.** The forest does not have water alone in the interspaces, there is usually thick mucilage so that the adventitious species in the silt and upper layer of mucilage do not penetrate downwards to the stone surface nor do the adnate species colonise the uppermost layers. The upright species are certainly divisible into a low storey (shrub-like) and a higher (tree-like) system, and the adventitious species live in the uppermost 'tree branches'. All these features mean that the epilithon, chosen for construction of indices by almost all workers, can vary from a simple 'ground flora' through to complex 'forest'—indeed, a most complex ecosystem has been selected. However, it is no more complex than the array of habitats in which the invertebrates live and as with the use of these, the fine detail has been obscured by the use of crude sampling techniques. Grazing by invertebrates certainly modifies the flora though most of the evidence shows that it is the upright species that are the most easily grazed leaving the adnate species, eg Hill & Knight (1987), De Nicola *et al* (1990). The remaining diatoms will still provide the indicators of water quality but grazing selection could be a disturbing factor if indices are computed. The effects of any patchiness caused by grazing can be overcome by sampling a sufficient area. From this discussion it will be clear that presence or absence of a species at any one site is not *solely* a function of water quality at that site. The most complex situation has been outlined because in designing sampling programmes and when interpreting data, it is essential to appreciate the possible complexity.

It should be noted that the adnate and pedunculate floras are, as the names imply, comprised of attached, non-motile organisms, whereas the adventitious, silt flora contains motile species. There are a few motile species also present in some sites within the mucilage but it is rare for these to be dominant. The flora in the mucilage has been termed the metaphyton in similar spatial/mucilaginous situations in lakes, Aloi, Lock & Goldman (1988) and in the comparable complex around the stems of aquatic plants, Behre (1966).

2.4 Sources of taxonomic information

Great advances in the classification of diatoms have been made in recent decades by use of scanning electron microscope studies. Older floras, eg Hustedt (1927–66, 1930), Patrick & Reimer (1966, 1975) should be used with caution since many taxa have now been re-assessed. The most up-to-date source is Krammer and Lange-Bertalot (1986, 1988, 1991 and the final volume in press) for species, and Round *et al*, (1990) for genera. Volumes of Bibliotheca Diatomologica, eg Archibald (1983), Krammer & Lange-Bertalot (1985, 1986, 1988, 1989, 1991a and 1991b), Reichardt (1984) are valuable and the new journal *Diatom Research* is dealing with many taxonomic problems and general diatom studies. Useful illustrations can be found in the following papers—these are accompanied by photomicrographs, Hartmann (1983), Leclercq & Manquet (1987), Watanabe *et al* (1988). There is unfortunately no single publication dealing solely with river diatoms. Whilst species are fairly widespread, especially throughout the northern hemisphere, it is necessary to be extremely careful when identifying species from distant regions using European floras; they may appear to be identical, but there are some subtle variations. This problem is much more pronounced when dealing with tropical and southern hemisphere samples. Many workers have simply identified to the nearest form in a European flora, and whilst this may lead to a perfectly correct identification, greater care is required than when using the flora in the geographical region for which it is designed.

Figs 4, 5 give examples of the common epilithic diatoms in British rivers.

2.5 Sources of ecological information in floras

Brief notes on the ecology of species are given in most floras, but they should be treated with extreme caution. The most reliable comments relating to occurrence in waters, eg xenosaprobic, oligosaprobic, α and β-mesosaprobic, polysaprobic, are to be found in the volumes of Krammer & Lange-Bertalot. Particular care is needed where species are described as occurring in nordic-alpine habitats—this usually means that the species was first found there. Few floras specifically refer to diatom distribution in rivers and the quoted ecological data usually refer to their occurrence in lakes—though without

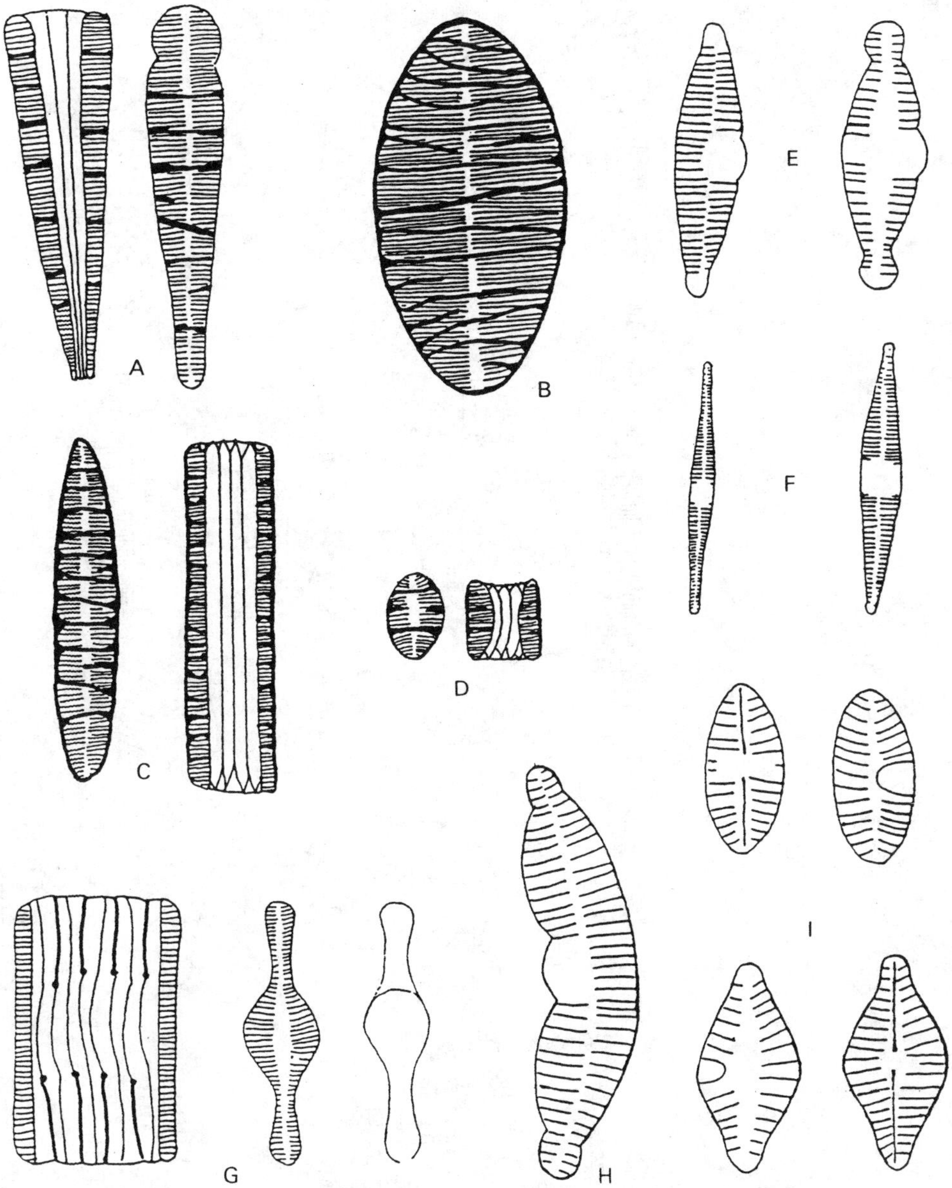

Fig 4. Common diatom species in the river epilithon: A *Meridion circulare*, B, C *Diatoma vulgare*, D *Diatoma mesodon*, E *Fragilaria vaucheriae*, F *Fragilaria capucina*, G *Tabellaria flocculosa*—a girdle view, valve view and girdle band, H *Hannaea arcus*, I Two forms of *Achnanthes lanceolata*—note the difference in the two valves of each form.

specifying this. There are several compilations of data on diatom tolerance to various water qualities, eg Lowe (1974), Bursik (1973), Descy & Coste (1987, 1988, 1990), Leclercq & Manquet (1987). These are extracted from the literature and further work is required to refine them since this method tends to stretch the limits of the individual species.

2.6 Introduced substrata

These were developed to provide standardised surfaces for colonisation by diatoms. Growth on these (eg glass plates, microscope slides, plastic plates, plastic tubing, nylon thread, cleaned sections of cut rock, etc) were considered by some workers to be easier than natural surfaces to sample. Colonisation is usually rapid and submersion for periods of 2–4 weeks is often adequate. Lamberti & Resh (1985) report good growth in 4 weeks and Biggs (1988) also considers 4 weeks to approximate the state of natural substrata in some rivers but up to 8 weeks in moderately enriched rivers—standard times were not recommended. Claims that these artificial surfaces provide data on all the

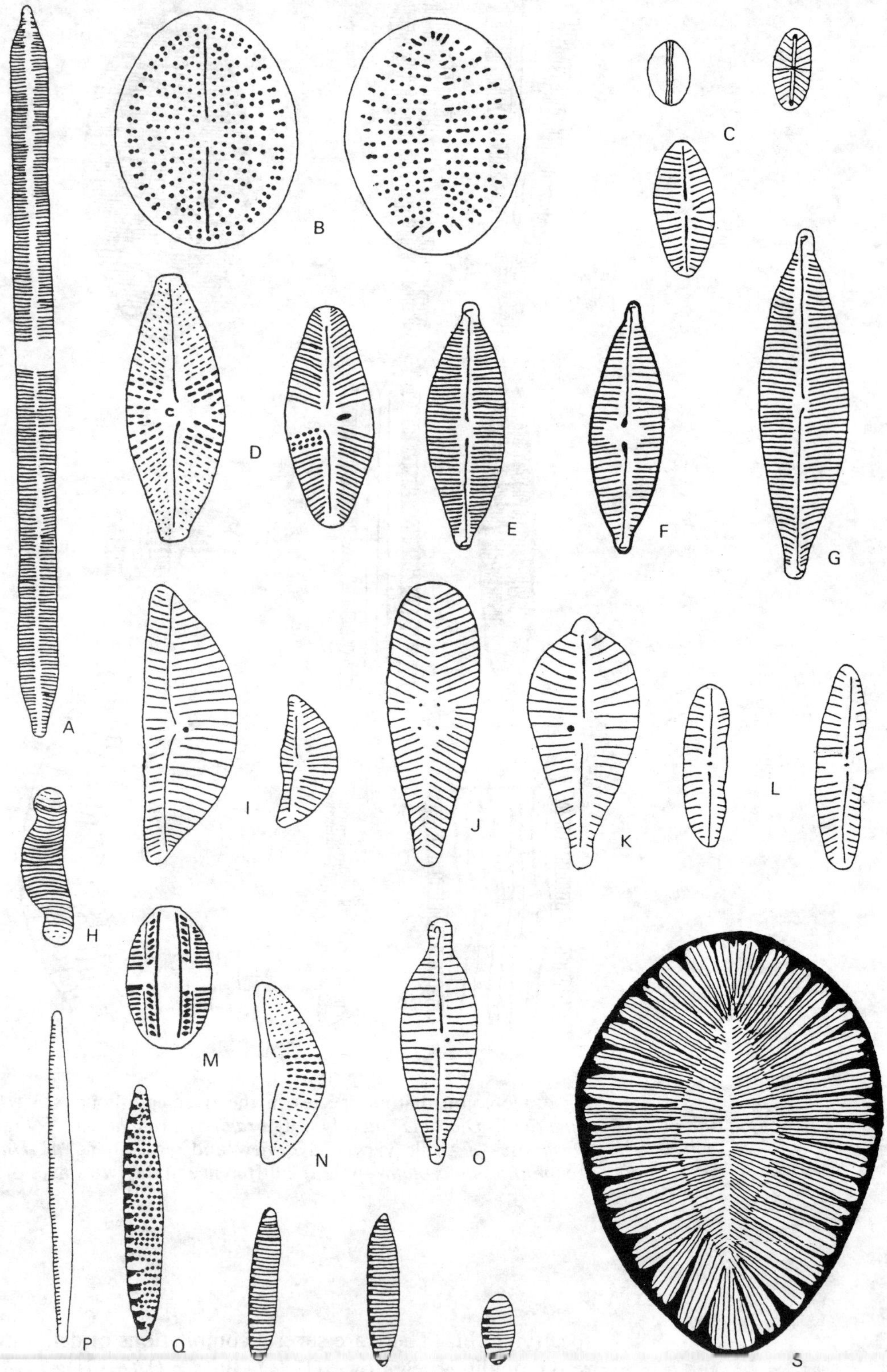

Fig 5. Further common diatom species in the river epilithon. A *Synedra ulna*, B *Cocconeis placentula*, C Three small *Navicula* species, D *Navicula geoppertiana*, E *Navicula accomoda*, F *Navicula halophila*, G *Navicula* sp., H *Eunotia exigua*, I *Cymbella minuta*, J *Gomphonema olivaceioides*, K *Gomphonema augur*, L *Reimeria sinuata*, M *Amphora pediculus*—girdle view, N *Amphora veneta*—valve view, O *Gomphonema parvulum*, P *Nitzschia palea*, Q *Nitzschia amphibia*, R *Nitzschia frustula*, S *Surirella ovata*.

species encountered on natural substrata have often been disputed and the arguments will no doubt continue. They can, of course, only represent the flora which is capable of developing from the inoculum present during the immersion period and not the integrated flora developed on natural surfaces over the preceding period (probably several months) of fluctuating environmental conditions. The advantages are in the ease of sampling from a smooth surface (but in many ways this is an unnatural surface), an easier control over exact area sampling and standardisation of substrata, less contamination from macrophytic growths, and they can be positioned exactly for surveillance of inflows etc where natural substratra may not be available. The disadvantages are that the flora is bound to be somewhat unnatural and biased towards those diatoms which are fast growing and can attach to flat, smooth surfaces (ground glass slides are sometimes used), the flora is not a 'climax' community, the smooth surface often leads to 'sloughing off' of the diatom film, apparatus has to be devised to hold the substrata to the stream bed, they are often lost or vandalised and they cost money to construct and place in the river. In addition sites have to be selected, whereas sampling stone surfaces is usually possible along the length of a river. The apparatus needs to remain in the river for 2–4 weeks so there is a delay before any information is available. Sampling the stone surfaces on the other hand can be completed and the counts of diatoms made in a few hours.

3 The method

See section about this booklet.

3.1 Method using epilithic diatoms for monitoring changes in river water quality

Field

3.1.1 Select a site where water is flowing over stones (preferably a riffle) which can be easily sampled. Where possible avoid disturbed sites (eg where cattle drink) or where there is extensive macrophyte cover.

3.1.2 Remove 5 small (up 5–10 cm diameter) preferably flat stones from the river bed in a suitable site away from the bank and at a depth under eg 0.5 m. Avoid stones covered with a coating of green algae or silt. Clearly it is not always possible to conform to this pattern and a degree of judgement has to be made at each site. Preliminary sampling and examination in the laboratory (see below) will rapidly enable workers to make good collections. Rich diatom growths are often visible on stones as a light/dark brown coating. In the absence of such a visible coating, the feel of mucilage on the stone surface usually indicates the presence of diatoms. No brown coating or mucilage is almost always due to recently inwashed or disturbed stones which have not had sufficient time to be colonised.

3.1.3 Place the stones (upperside uppermost) in a suitable container (plastic box, etc) which will prevent drying out and preferably keep in the dark.

Laboratory

3.1.4 Remove the flora from an area of the upper surface of each stone by rubbing with a finger, brushing (a toothbrush is suitable) or scraping with a scalpel, etc and washing the loosened diatoms into a beaker or centrifuge tube with a jet of distilled water. Simple as it is, rubbing with a finger is often the best technique, since by feel the complete removal of the mucilage coating can be detected. Brushing or scraping often removes inorganic fragments which merely obscure the diatoms in the later microscopic study*. If very detailed quantitative data are required, see Appendix 7.

3.1.5 Allow the sample to settle overnight or centrifuge and use immediately. Pour off or remove excess water via a narrowed glass tube with upturned tip connected to a vacuum device on a water tap. The amount left is not critical (0.5 ml of sediment/water mixture is plenty).

---

* At this stage, a subsample can be observed under the microscope to check that live cells are present, or dead cells left by grazing organisms, or the degree of growth of other algae etc—many useful insights into the complex community can be gained but it is not essential for actual monitoring of the diatom populations.

3.1.6 Add a volume (2 ml) of saturated potassium permangenate solution to the sediment, shake well and leave overnight or proceed immediately to the next stage.

3.1.7 In a fume chamber, add an equal amount of concentrated hydrochloric acid, cover the tube or beaker with cling-film and place in an oven (or water bath or any heating device) at 60°C until the liquid turns almost colourless (a slight yellowish clear colour).

3.1.8 Either settle for 24 h and then centrifuge or centrifuge when cool. Remove the supernatant, resuspend in distilled water and repeat 5/6 times. Shake the tube between each centrifugation in order to separate the pellet of diatoms and allow water to flow around the material to remove acid from the surfaces. This washing is necessary to remove all traces of acid, 5/6 centrifugations is usually sufficient (see below). NB. If pouring off the supernatant, do so in one action, inverting the tube once, the diatom pellet at the bottom will remain in the tube. Suction can be via a glass tube with a turned up tip as used in the settling technique for phytoplankton.

3.1.9 After the last centrifugation, the sample will appear greyish and cloudy if a large population of diatoms has been cleaned. If almost colourless, this may indicate a small population on the original stone surface or loss of cells during preparation. It is usually necessary to dilute the sample considerably with distilled water so that the material placed on a coverglass will give an even spread of diatoms. If the valves pile on top of one another, simply dilute further.

3.1.10 Shake the sample well to mix the diatoms and place a drop of suspension onto a coverglass, leave to dry (or heat gently on a hot-plate; this may leave the diatoms clustered but experience will determine the technique to use) and then invert the coverglass onto a drop of mounting medium. Naphrax is a good medium* but others, eg Hyran can be used. Canada Balsam should be avoided, since, if high resolution microscope lenses are used, a high refractive index mountant is essential. Leave the slide on a warm hot-plate to harden—again, experience will yield the time from a few minutes upwards depending on the temperature. **NB.** Do not dry a sample onto a microscope slide and then add mountant and coverglass. The diatoms are best viewed without looking through mountant—hence the preference for drying onto the coverglass.

3.1.11 Label the slide with details of sample number, site, date, etc. Stick-on labels can be used. However, these can come off or the writing fade, so it is good practice to number the slide with a diamond marker.

Counting

The slides will then be ready for determination of species and counting**, which is preferably done using a × 10 eyepiece and × 100 oil immersion objective. Counting can most simply be done by selecting fields at random and counting cells until 100 have been recorded. Counting of more than 100 will only yield rare and often contaminating species since it must be remembered that diatoms colonise all available micro-habitats in the river and in addition species can be washed in from tributaries, ponds, reservoirs, lakes, soils, in all of which diatoms are abundant. These may appear to be a problem but they are not, except when large numbers of cells are counted. Broken valves should be included only when more than half the valve is present. Girdle bands should not be counted. Some cells remain intact and these will certainly bias the counts. If all cells of a species are intact then each has two valves and can be recorded as such. However, owing to the lie of cells, it may not be easy to ascertain whether one or two valves are present. The dominants in samples are usually so obvious that this problem is rarely important for the present purposes. Equally, there will always be a small proportion of cells which for one reason or another cannot be identified.

When a site is to be visited and sampled on a regular basis, it is useful to compile a table of the common species and duplicate this for recording (see Appendix 1)—in this way changes in the flora over time will be obvious.

---

* Available from Northern Biological Supplies, Ltd., 3 Betts Avenue, Martlesham Heath, Ipswich.

** Sections 3.1.4–3.1.11 can be accomplished within 2 h or done in stages over several days.

The commonest species and those at present of greatest indicator value are illustrated in Figs 4, 5 and have been included on the recording sheet (Appendix 1). Photographs of examples of material prepared as above are provided in Appendices 2 (light micrographs) and 3 (scanning electron micrographs). As can be seen from these, the abundant species tend to be dominant in a single microscope field.

For further useful information it is suggested that sections 2, 3 and 5 be studied. This is especially important if the worker has little previous experience of sampling such communities.

3.2 Additional comments

Methods will depend very largely on the object of the exercise, precision required, time for the analysis and length of river. Provided the problems are understood, sampling can be varied but should be consistent in any one programme and the details of sampling should be reported in detail. It may be necessary to perform a preliminary survey to determine the overall features of the available habitats. In all surveys, the degree of precision required should be established—multiple sampling at some sites should indicate how many samples need to be taken and how many microscopic observations made. All will depend on the object of the exercise.

Apparatus and chemicals required

1. Plastic boxes for stone samples
2. Beakers or centrifuge tubes
3. Cling film
4. Fume chamber
5. Water bath or simple oven to heat to 60°C
6. Simple centrifuge. Machines without braking devices are preferable since the latter often cause mixing and loss of diatoms during the slowing down
7. Compound microscope, equipped with $\times 100$ oil immersion lens
8. Microscope slides and cover glasses, slide labels, slide boxes
9. Naphrax (or other high refractive index) mounting medium
10. Simple pipettes/rubber teats
11. Hot plate
12. Distilled water. (Do not use tap water which can contain a few diatoms)
13. Concentrated hydrochloric acid.
14. Saturated potassium permanganate solution. (Keep in a foil covered bottle).

3.2.1 Sampling—Field

3.2.1.1 Selection of site

It is usual to suggest random sampling for ecological purposes but for the present work some selectivity is essential to avoid unnecessary work. For whole river surveys, sites should be selected at suitable intervals to include each characteristic stretch and each geological/agricultural/urban section. Stations may have to be close together especially in the upper reaches (rhithron) where changes in water quality are rapid—in the lower reaches (potamon) there is much greater stability and stations may be more distant; in the lowest reach the estuarine effects need to be considered. Clearly the choice will depend on the object of the exercise. It might be advisable to make a decision (if possible) to avoid shaded sites—McIntire (1968) showed that light intensity had an effect on diatom associations in laboratory streams though the extent in natural situations needs more investigation. Robinson & Rushforth (1987) and Hill & Knight (1987) report more species at open sites than at shaded sites although most of the common species were present at both; Winterbourne (1990) found no difference in the dominants in shaded and open streams in New Zealand. Biomass is certainly higher at high irradiance levels, Steinman & McIntyre (1987)—they report the adnate species dominant at low and the rosettes/filaments at high irradiance.

Many tables in the literature can be confusing because data from side-streams are entered and these may be from waters of totally different quality to that of the main stream. They should be kept separate.

On the whole, all species are present throughout the year but slight season variations have been recorded, eg Engelberg (1987) considers there are two seasons—winter/spring and early summer/autumn although most diatoms are present in both. There are confusing statements in the literature because there can be massive increases in biomass (cell numbers) in the spring but in general the indicator species are present at all seasons (though cell counts of individuals will always be changing).

3.2.1.2 Selection of material

In the upper reaches it may be necessary to sample bedrock but in most cases it is possible to find stones. In all other situations it is best to select stones from the riffle regions where they are subject to good water flow and therefore not grossly contaminated by silting. In the potamon (the slow flowing stretch) this is not always possible and any available stones may have to be taken and the silting problem dealt with (see 3.1.2). Most workers recommend collecting stones (circa 5–10 cm diam.) with flat upper surfaces. Five stones are usually adequate, taken at intervals rather than adjacent. Again, selection makes sense and stones obviously coated with brown mucilage will be rich in diatoms. Occasionally brown colonised stones are sampled only to find that the brown is due to coloration of mucilaginous sheaths of filamentous Cyanobacteria. Dussart (1982) comments on the waste of time if newly washed-in stones are collected. In some sites, if stones do not have a visible coating of diatoms, then stones with a mucilaginous film on the surface are most likely to support an epilithon—stones with no mucilage coating have usually not had time for a mature flora to develop. Stones bearing macroscopic growths of algae (*Lemanea*, *Batrachospermum*, *Cladophora*), mosses or liverwort should be avoided (see 3.1.2). Where possible it is best to select stones of the same common rock type at each sampling site—this may reduce the variability in the counts and provide the flora best representative of the water quality at that point.

Place the stones in labelled polythene boxes without adding water. For general surveys, there is no loss of data if stones touch each other, but for extremely accurate cell counting, studies of variation in the flora on different faces of stones, on stones of differing rock type, etc, greater care must be taken and the individual stones transported carefully.

3.2.2 Sampling—Laboratory

Techniques will again vary according to the object of the exercise but over-riding consideration has to be the selection of relatively* uncontaminated surfaces for sampling. Occasionally stones cannot be used because it is not clear which is the upper surface—the lower surface can retain old, dead cells and even have enhanced numbers of cells of species adapted to the more anaerobic under-stone surface thus providing false information. Assuming that the field sampling avoided stones with macroscopic growths, it is now necessary to inspect the surfaces more carefully and choose areas of stones where there are no young growths of macroscopic plants or other contaminants. Since the diatoms are microscopic and almost always in layers on the stone surface (the epilithon is actually more like a forest structure condensed down to a few millimetres height—see Fig 3), direct microscopic observations are rarely possible—but see below.

Ideally each component of the 'forest' should be sampled separately but this is not necessary for most water quality studies. Indeed they can be mixed (as most workers do) provided one is aware of their existence and composition when interpreting data. The one which is most easily recognisable and removable is the adventitious 'silt' flora. Gentle washing of the surface, with a jet of distilled water—coupled with light brushing with a brush or finger—will remove most (but certainly not all) cells of this community. Table 1 gives counts of the dominants using this technique.

* It is sometimes impossible to find stones which have no germlings of macroscopic algae/mosses/liverworts, but usually the epilithon is sufficiently rich to obscure any epiphyte contaminants (see also 3.1.2).

Table 1 Two examples of counts (%) of the three dominants on unwashed (ie silted) and washed (ie silt washed off)

R. Hodder			R. Aire		
unwashed	*Navicula lanceolata*	36	unwashed	*Navicula lanceolata*	53
	Achnanthes linearis	28		*Amphora pediculus*	21
	Navicula gregaria	16		*Navicula gregaria*	12
washed	*Cymbella sinuata*	45	washed	*Amphora pediculus*	62
	Achnanthes linearis	14		*Cocconeis placentula*	11
	Cymbella minuta	34		*Cymbella minuta*	12

The sample removed from the stone surface will be of live/dead/broken cells and it is always preferable to check this under the light microscope for proportions of live/dead cells. Usually the majority are live but during periods of intense grazing (Protozoa can consume enormous numbers, as can invertebrate larvae, etc) almost all cells (especially of the silt flora) can be dead, voided cells. Cell counts could be made at this stage, but there is often so much detritus present that accuracy in both identification and counting is difficult. Counting of live diatom cells has been performed in a very few studies, eg Leland & Carter (1984), Steinman & Lamberti (1988). Almost all workers use standard techniques to remove organic matter from the cells (and detritus). The simplest techniques involve cold or hot oxidation, followed by repeated washing and settling (usually centrifuging) in 5–6 changes of distilled water. The cells in the distilled water are then thoroughly mixed (shaken) and a sub-sample pipetted onto a cover glass, dried (overnight or quickly on a hot plate) and then mounted on a microscope slide in a high resolution mounting medium. If any acid or acid/water mixture remains at the mounting stage, each valve will be enclosed in a bubble and determination of the species will not be possible.

3.2.3 Cell counting

Some form of quantitative estimation is usually essential even if the data are subsequently converted to numerical categories. Counting also involves identification of species (identification merely to genus is hardly ever sufficient for ecological purposes). It is essential to identify accurately but since the number of species in any one sample can be considerable (some Continental and Japanese studies involve between 100–300 species) some simplification is essential for routine studies. An illustrated guide to the commonest epilithic species is provided in Figs 4, 5 and Appendices 2, 3, and notes on species in Appendix 4.

3.2.4 Minimum numbers

The number of cells counted varies from 100 cells per sample, to the most commonly used 500 per sample, to the excessive counts of several thousands (in the glass slide field sampling technique) used by some American workers. The more cells counted, the greater the number of species recorded and this involves excessive time spent learning to identify rare species. One reason for the need for large counts is due to the fact that the initial sampling often encompasses the attached flora, the silt flora and frequently the epiphytic flora of small macroscopic species. A second reason is that some counting systems are related to diversity which does require large counts (however, diversity has been shown by many workers to be unsuitable for (water quality) monitoring, see Lange-Bertalot (1979) and references quoted). Care with the initial sampling and subsequent removal of the flora from the stone surface will reduce the need for large counts. This is simply because in well-defined, single micro-habitats it is usual to find only a small number of species but a high number of cells of these, ie niches are occupied by a few well-adapted species. Excessive counting adds many of the casual contaminant species and these are of minimal indicator value. In any case, study of the data in the literature reveals that in almost every instance, rare species (often defined as less than 1%, sometimes less than 5%, of the counts) are left out of the calculations or are recorded but ignored. It is, in fact, rare to find even as many as five species in *abundance* in a sample—one or two are the usual. Experience often shows that a single microscope field reveals the dominants (see Appendix 5) and a count of at most 100 cells is recommended. Some workers simply recommend a series of abundance classes (eg 1 = rare, up to 6 = very abundant, and Dussart (1982) has shown that data from such estimates, when plotted against a $\log_{10}$ of cell counts, give a linear relationship). Levandowsky (1972) also suggested that orders of magnitude (1–10, 10–100 etc) are sufficient, and Moller & Pankow (1981) suggest merely a scale of 1–5 or even 1–3 for abundance is adequate. If abundance classes are ultimately used, it is still better to base these on a simple count since this concentrates the attention of the worker on identification of the dominants. Numbers do not of course express biomass and this has only rarely been taken into account in river surveys; it is simple enough to compute from a combination of the counts with cell volumes if standard areas of stone surface are sampled. Biomass data would require a considerable re-assessment of indices which are based on cell counts and are not recommended here. In addition, biomass can bear little reationship to production since grazing often has a dramatic effect.

3.2.5 Counting errors

These may arise from several sources. The first is clumping of cells or uneven distribution on the coverglass when preparing slides. It can only be overcome by preparing new slides after thorough mixing of the cleaned sample. Addition of substances such as dilute surfactants have been proposed but it is not usually necessary. Incomplete washing of the sample leaving acid behind causes bubbles around cells which are then difficult to identify, necessitating return to the cleaned sample and again washing several times in distilled water. The drops of sample placed on coverglasses do not dry down evenly especially if heat is used—it is advisable to check evenness of the sample and usually avoid the edges. If extreme accuracy is required, the method devised by Battarbee (1973) for preparing slides can be used.

Since all diatoms consist of two valves and a number (often 4, but sometimes up to 50) of girdle bands, errors can arise since some cells stay as complete entities and others split into the two valves and girdle bands, see Round, Crawford & Mann (1990) for a detailed account of diatom morphology and generic features. The first rule is to ignore the girdle bands (see Fig 1) and count only valves. If cleaning has been vigorous, most cells will have split into the two valves but in a few species separation of the valves is difficult. Some valves (or cells) will lie in girdle view and it will be more difficult to assign these to species, though experience will often lead to recognition of species in girdle view. Valves sometimes fragment—a simple rule is to count only fragments where more than 50% of the valve is present. Finally, there are nearly always a few valves lying in positions where accurate identification is not possible. On most slides the above situations will be minimal and will only slightly modify the accuracy of the counts. Absolute accuracy is not attainable in any ecological study but errors will be kept to the minimum if reasonable care is taken at each stage. Many counts in the literature provide spurious accuracy and conclusions should only be drawn from the presence of clearly dominant forms.

3.2.6 Tabulation of counts

Preliminary studies should be undertaken to establish the dominant species in the sampling area. A standard form for recording each count can then be used. This has the advantage that any change will immediately be obvious and re-checking of the site or sample can be made if necessary. A suggested is provided (Appendix 1).

4 Indices

> 'any particular scale . . . may be objected to on grounds of bias, low precision, restricted generality and other factors but objectors should remember that these are relative and practical matters and that no scale used by mortals is perfectly free of their taint.'
>
> Stevens (1946). On the theory of scales of measurement, Science, **103**.

The purpose of indices is simply to reduce floristic (faunistic) data to a numerical form. This most frequently ends up as a series of numbers from 1–5 representing five classes of organisms which are then usually related to water quality (also often expressed by five classes). The indices are simple arithmetical expressions based on some form of counting of species and usually coupled with factors derived from the ecological tolerance of the species (sensitivity index to pollution) and often an 'indicating (sic) value' or estimate of each species 'ecological amplitude towards pollution' (Descy 1979). It is important to note that the use of diatoms (and other organisms) in devising indices is simply an application of the concept of 'indicator species communities', see comments in Hellawell (1972). Two essentials are necessary before an index can be calculated—the diatoms must be identified and counted (a standardised, illustrative, simple river flora is needed—and a list containing numerical estimates of the species sensitivity to pollution and indicator value, see Appendix 7). The concept of these indices goes back at least as far as Pantle & Buck (1952) with further data and estimates of 'reliability' etc in Zelinka & Marvan (1961), Fjerdingstad (1964, 1965), Sládecek (1973), Lowe (1974), Wegl (1983).

Sampling and counting is first performed and the indices are then calculated from the tabulated data; the following is a selection of indices which can be calculated from data produced by the techniques described in section 3.

4.1 Kothe's index

This is a very simple system based on a concept known as 'species deficiency' calculated from the formula

$$\text{species deficiency index} = F = \frac{A_1 - A_x}{A_1} \times 100$$

where A_1 is the number of cells of a species chosen as a standard at sampling site 1 and A_x is the number of cells of that species at site x. The number of diatom cells (valves) to be used can be worked out by calculating the index for the first 50 valves, then the second 50 and so on, plotting the indices against the cumulative total of valves, ie, 50, 100, and so on, and where the index curve levels out is the correct number to use. In the case of river quality it could only be used by selecting different species in a sequence down a river since none of the species of extremely pure water occur elsewhere, eg *Eunotia exigua* could be the first species and the formula would give the decline in this species as the water conditions deteriorated. But it would be absent from most of the length of eutrophic rivers (the majority in the UK). It is a simple technique which has been used only rarely, see Economou-Amilli (1980) and Wegl (1983). The idea of providing a measure based on deviation of the community from that considered typical for a river zone has also been suggested by Hawkes (1982). Ramm (1988) has a similar technique comparing fish populations against the potential numbers of taxa in a clean habitat, see also Hellawell (1972) for a discussion of various types of indices.

4.2 Crains' sequential comparison index

This again is simple to use and has the great advantage that no species identification is necessary. When used for diatoms, all that is necessary is to traverse a slide using some form of marker in the eyepiece and each time the marker meets a diatom it is recorded. The first diatom can be designated by any code, eg 1 or A, the next by 2 or B, and so on. If the sequence runs 1, 2, 1, 2, 1, 2 then there are six runs each involving two species. If the sequence is 1, 1, 1, 2, 1, 1, 1 then there are three runs involving the two species. The index is calculated by dividing the number of runs by the number of species encountered, thus in the first example, it is $6/6 = 1$, in the second $3/7 = 0.43$. This system was tried by Archibald (1972) who found it was not entirely satisfactory but as effective as other measures. These first two methods need testing, though the sequential comparison index has been successfully used by Hodgkiss & Law (1985).

4.3 Descy's index

This has been one of the most frequently quoted. It is based on Descy (1979, 1980). The counts are used to calculate a 'diatom index' (Id) from the following formula.

$$\text{diatom index} = \text{Id} = \frac{\sum_{j=1}^{n} A_j \times i_j \times v_j}{\sum_{j=1}^{n} A_j \times v_j}$$

To avoid multiplicity of symbols, Descy's notation is followed where A_j is the relative abundance of species j in the sample, i_j the sensitivity of species j and v_j the indicator (reliability) value of species j and n equals the number of species counted in the sample.

Since i_j values range from 1–5 and v_j from 1–3 this calculation will give values ranging from 1–5. Descy (1979) then relates the values to water quality thus:

Id > 4.5: Best biological quality, no pollution

Id 4.0–4.5: Almost normal quality (slight changes in the community, slight pollution)

Id 3.0–4.0: More important changes in the community, decrease of the sensitive species, moderate pollution or significant eutrophication.

Id 2.0–3.0: Resistant (to pollution) species dominant, decrease or disappearance of the sensitive species (reduced diversity), heavy pollution.

Id 1.0–2.0: Marked dominance of a few resistant species (many species have disappeared), very heavy pollution.

An example quoted in several publications is from a site heavily polluted by domestic sewage in the Belgian River Semois in which the floral composition was *Achnanthes lanceolata* 0.9%, *Gomphonema parvulum* 0.5%, *Navicula accomoda* 78.7%, *Nitzschia palea* 19.7% and others 0.2%. The method of calculation is as follows:—

Achnanthes lanceolata
0.9% ($v_j = 1$) (product = 0.9) ($i_j = e$) (product = 2.7)

Gomphonema parvulum
0.5% ($v_j = 1$) (product = 0.5) ($i_j = 2$) (product = 1.0)

Navicula accomoda
78.7% ($v_j = 3$) (product = 235.1) ($i_j = 1$) (product = 235.1)

Nitzschia palea
19.7% ($v_j = 2$) (product = 39.4) ($i_j = 1$) (product = 39.4)

Others
0.2% ($\Sigma = 275.9$)–($\Sigma = 278.2$)

Id = 278.2/275.9 = 1.0 ie heavy pollution

The Descy index has been criticised by Leclercq & Manquet (1987) since it results in high values and thus the under-estimation of moderate/heavy pollution. This is due to the use of too few species and the absence of the small *Navicula* spp. from the indices. Descy (1979) also provided some graphical examples of diatom distributions, two of which are provided here (Figs 6, 7).

4.4 Coste's index CEE 88

This system originally devised for work on the R. Seine, Coste (1976) and amplified later, Coste (1984), Descy & Coste (1990) in the 1984 form based on 11 species, and in the 1990 form 202 species, is displayed in a grid comprising eight groups (G1–G8) of species, the first (G1) which is characteristic of clean water ranging through to G8 which are characteristic of polluted water, and four subgroups of species of more euryoecious nature, ie having a wide tolerance, but broadly representative of clean acidic or alkaline waters (SG1), through increasing alkalinity/mineralisation to the (SG4) group which occur in slightly saline waters (Table 2). Using the example from Descy (see above) the

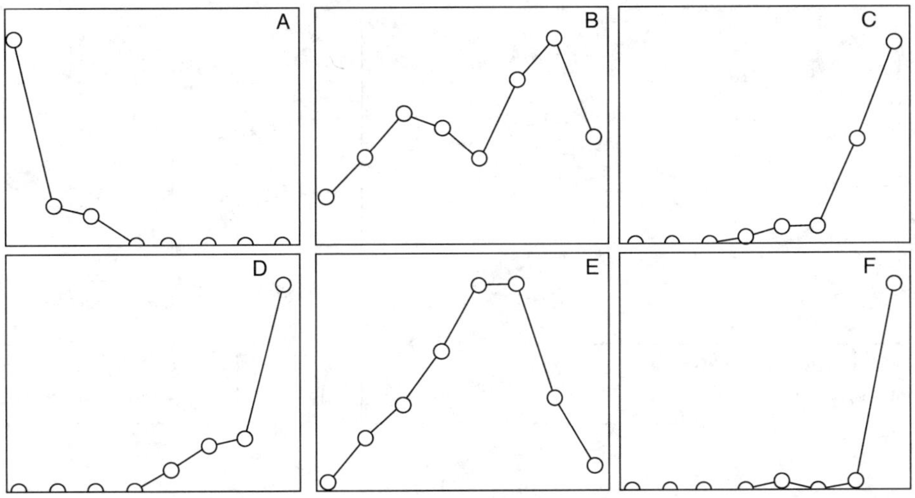

A: *Navicula accomoda*, B: *Achnanthes minutissima*, C: *Navicula menisculus*, D: *Achnanthes microcephala*, E: *Navicula cryptocephala*, F: *Eunotia exigua*.

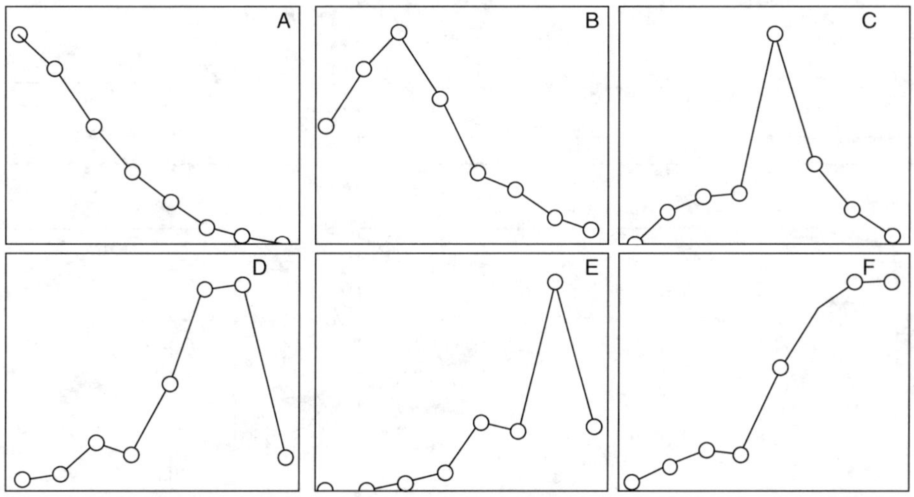

A: *Nitzschia palea*, B: *Gomphonema parvulum*, C: *Nitzschia paleacea*, D: *Cymbella ventricosa*, E: Nitzschia fonticola, F: *Fragilaria capucina*

Fig 6. Some examples of diatom abundance relative to water pollution—greatest on left (from Descy)

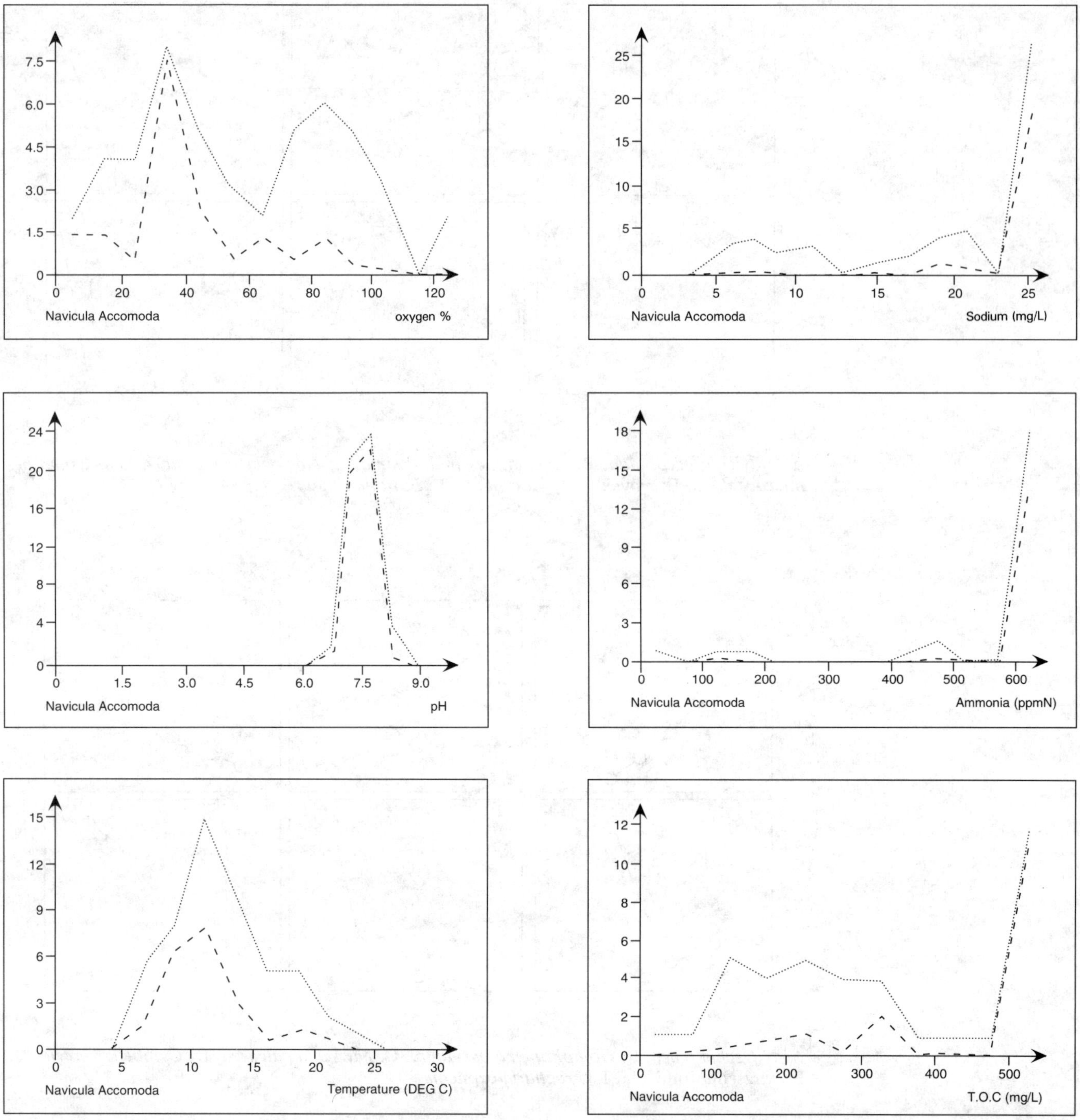

Fig 7. Occurrence of the diatom *Navicula accomoda* along environmental gradients (from Descy).

SG1	AAUS	ABIO	ACLE	AEXI	AFLE	APEL	APUS	AVIT	CCES	CEHR	CGRA	CSHU	CSLE	DANC
	DHME	EARC	ECUR	EEXI	ERHO	FRSA	FVIR	GCLA	GCLE	NACD	NCLE	NEAF	NEAL	NGPE
	NHAN	NOBL	NPSL	NRAD	NRHE	NRHY	NSPD	NSTL	NTRI	NTUS	PNOB	RGIB	SLIN	STAN
SG2	AMAR	ANOR	CAFF	CBAC	CCIS	CCYM	CELL	CHEL	CLAN	CMIC	CNAV	CPED	CPLA	CSIL
	CSTE	CTGL	CYLI	DOBL	ETUR	FCON	FCRO	FVUL	GAFF	GGRA	GNOD	GOMI	GTRU	GYAC
	NACU	NCTE	NINO	NMEN	NREC	NSBH	NSBL	NTPT	NVER	PGIB	PSCA	PVIR	SACU	SANG
SG3	ACON	AEXG	ASAX	AUGA	AUGR	CDUB	CPST	FBRE	FLEP	FPIN	GANG	GAUG	GTER	MAMB
	NCPR	NEXI	NFIL	NGRE	NHEU	NIAN	NIGF	NIGR	NIRO	NITE	NKOT	NLAT	NLIN	NPEL
	NSLE	NTRD	NTRL	NVIR	NVRO	RLON	SHAN	SIDE	SPAR	SPUL	SROT	STKR	SULN	SUOX
SG4	ADEL	ASPH	BPAX	CAMP	CATO	CCAS	CMEN	CSTR	GGLI	GSCA	NALE	NAMC	NAMP	NCIN
	NCLA	NCUS	NDUB	NHAL	NIFR	NIFT	NINT	NIPU	NMUT	NMVE	NOBT	NPHY	NPRO	NPUP
	NSHR	NPYG	NRCS	NSAL	NSIG	NTEN	NTRV	NTRY	PMIC	SOVI	SPAV	STAB	STTU	TBRA

	G1	**G2**	**G3**	**G4**	**G5**	**G6**	**G7**	**G8**
	AMIN	ALIB	ALAN	FCVA	AMMO	GPAR	AVEN	NGAN
	DTCR	AOVA	AROS	MVAR	DITE	NMIN	NACO	NGAT
	EPEC	APED	CCAE	NJOU	NACI	NATO	NCOM	NPAL
	GANT	CMIN	CPRO	NLAN	NAPI	NIAR	NGOE	NPAD
	HARC	DVUL	CSIN	NROM	NIHU	NMLF	NSEM	NZSU
	MCIR	FCAP	GMIN	NSOC	NPAE	NMMU	NSMO	NUMB
	NSIN	NDIS	GOLI	RABB	SOVA	NSBM	NVEN	
	G1	**G2**	**G3**	**G4**	**G5**	**G6**	**G7**	**G8**
SG1	10	9	8	7	6	5	4	3
SG2	9	8	7	6	5	4	3	2
SG3	8	7	6	5	4	3	2	1
SG4	7	6	5	4	3	2	1	0

Table 2 A grid as used by Coste for his method to determine the diatom index of water quality. Species are abbreviated, eg in G1, AMIN = *Achnanthes minutissima*; **DTCR =** *Denticula tenuis* var. *crassula*; **EPEC =** *Eunotia pectinalis*; **GANT—** *Gomphonema angustatum*, etc.

dominants are *G. parvulum* (Group 4) 0.5%, and *Navicula accomoda*—78.7% and *Nitzschia palea* 19.7% (Group 5)—total 98.4%. The diatom index is obtained from the formula (12–2GM). GM designates the median of the group, which in this case is Group 5 (ie the group which contains the diatoms with the highest count, that is *N. accomoda*). Therefore, $12-2\times5=$ (diatom index 2 which indicates a high degree of pollution. Taking another example, Coste (1990) from the Ardèche (see Table 3) where there are counts which fall in both the groups and sub-groups, the index is calculated from $12-(SGM=GM)$. The median figure of the sub-groups (53) falls in SG5 and the median figure of the groups (354) falls in G2; the diatom index is therefore 5. This method of calculating an index of pollution is now the favoured one after the exhaustive work of Descy & Coste (1990).

Table 3 An example of the calculation of the Coste index, from Coste (1990)

Results of a calculation based on data from the Ardèche on 10.07.84

	%	Groups or sub-groups	
Achanthes lanceolata	2	5	
Achnanthes linearis	1	2	G1
Achnanthes minutissima	38	91	G1
Amphora pediculus	1	2	G3
Cymbella affinis	1	2	SG2
Ceratoneis (Hannaea) arcus	1	2	G1
Cyclotella meneghiniana	1	2	
Cocconeis placentula	4	10	
Cymbella sinuata	6	4	G2
Cymbella ventricosa	41	98	G2
Gomphonema dichotomum	1	2	G1
Gomphonema parvulum	24	58	G4
Melosira varians	3	7	G3
Nitzschia amphibia	16	38	G5
Navicula bacillum	1	2	
Navicula exigua	4	10	SG3
Navicula frugalis	16	38	G4
Navicula gregaria	1	2	G3
Nitzschia hantzschiana	2	5	SG1
Navicula ignota	7	17	SG4
Navicula menisculus	1	2	SG3
Nitzschia paleacea	24	58	G4
Nitzchia palea	85	204	G5
Navicula pupula	22	53	SG5
Nitzschia romana	101	242	G2
Nitzschia perminuta	3	7	SG3
Navicula tantula	3	7	G3
Nitzschia umbonata	1	2	G5

Sub-groups:

	Total	Cummulative		Total	Cummulative
SG1 = 5	5	5	G1 = 2 + 91 + 2 + 2 =	97	97
SG2 = 2	2	7	G2 = 14 + 98 + 242 =	354	451
SG3 = 10 + 2 + 7	19	26	G3 = 2 + 7 + 2 + 7 =	18	469
SG4 = 17	17	43	G4 = 58 + 38 + 58 =	154	623
SG5 = 53	53	96	G5 = 38 + 204 + 2 =	244	867
TOTAL SG:	96		TOTAL GROUPS:	867	
Median	48		Median 433.5		

SGM = 5 GM = 2 diatom index = 5

4.5 Leclercq & Manquet's index (1987)

This is a more detailed index (using over 400 species) similar to Descy's, but with the values to be entered into the calculation based on a more detailed compilation from the literature and giving each diatom a saprobic valency from 1–5 in 0.5 intervals and indicator (reliability) values from 1 to 5 (Descy used valency values 1–5 in intervals of 1 and indicator values 1–3). Such a large number of species is difficult even for diatom experts to handle and so the system could never be widely used. Again, the major complication is the inclusion of many species which are either epiphytic or epipelic.

4.6 Lange-Bertalot's index (1978, 1979)

Lange-Bertalot worked for several years on the diatoms of the R. Main and other European rivers. The R. Main is one of the most, if not **the** most, polluted rivers studied in Europe and hence his perspective was from pollution to clean water and his concept of tolerance is towards the polluted water and never towards clean, ie when he writes of tolerant species, he means 'tolerant of pollution'. He has four water classes but does not list species in Class I waters (quite unpolluted). The Lange-Bertalot (1979) system is perhaps the simplest, since the author classifies the diatoms into five groups—Group 1, Extremely tolerant taxa. Dominant in extreme (polysaprobic) sites. Group 2, Tolerant taxa—divided into two subgroups. Group 3. Sensitive taxa (ie sensitive to pollution)—divided into two subgroups. Calculating the Lange-Bertalot index for the species in Descy's example is simple since 98.4% of the cells in the count are taxa in Group I giving a water quality assessment IV or polysaprobic. Kobayasi & Mayama (1982) found that the Lange-Bertalot system worked well on a polluted Japanese river. Steinberg & Schiefele (1988) expanded this system to take into account the degree of nutrient loading as well as pollution and produced a system leading to seven water quality classes using the proportions of seven groups of diatoms. This worked well on a river with several inputs from sewage works and a fish farm.

4.7 Watanabe's Index (DAIpo)

The system used by Watanabe and co-workers in Japan was devised in relation to organic pollution of river waters. The first publication was in 1981 in Japanese and the first in English was Watanabe (1986): see also Watanabe, Asai, Houki (1988) and there are more than 25 publications using these indices. This work is probably the most detailed on rivers ever made; it has also been extended to lakes and used by a number of Japanese workers in many laboratories. He considers that physical factors, in addition to chemical, affect the flora and hence suggests that collections be of rich growths, from the upper surfaces of flat stones, collected from riffles about 30 cm below the water surface and with flow rate approximately 40 cm/sec. These workers identify (in the 1986 paper), 454 diatom species, obtained from 65 rivers, but in the 1988 paper there is a table using only 65 species which is reproduced here (Fig 8) showing the distribution of major species relative to the DAIpo index—this is a good illustration of the spread of diatoms through waters from polluted to clean. The DAIpo value is calculated from either of the two equations:

$$\mathrm{DAI}_{\mathrm{po}} = 100 - \sum_{i=1}^{m} \mathrm{S}_i - 1/2 \sum_{j=1}^{m} \mathrm{E}_j \qquad (1)$$

or

$$\mathrm{DAI}_{\mathrm{po}} = 1/2\,(100 - \sum_{i=1}^{m} \mathrm{S}_i - \sum_{k=1}^{p} \mathrm{X}_k) + \sum_{k=1}^{p} \mathrm{X}_k$$

$$= 50 + 1/2\,(\sum_{k=1}^{p} \mathrm{X}_k - \sum_{i=1}^{m} \mathrm{S}_i) \qquad (2)$$

where

$\sum_{i=1}^{m} \mathrm{S}_i$: sum of relative abundances (%) of saprophilous taxa from 1 to m;

$\sum_{j=1}^{n} \mathrm{E}_j$: sum of relative abundances (%) of eurysaprobic taxa from 1 to n;

$\sum_{k=1}^{p} \mathrm{X}_k$: sum of relative abundances (%) of saxproxenous taxa from 1 to p.

The information in the Japanese papers provides a most comprehensive collection of river epilithic diatom distribution data in relation to organic pollution, pH and conductivity.

4.8 Palmer's index

This is a list of diatoms (and other algae) compiled from the literature; each alga is given a rating for its tolerance ot organic pollution, 1 = intolerant, 5 = tolerant (Palmer 1969). Whilst the diatoms more or less fall into the expected sequence (eg selecting those species which are associated with stone surfaces, the sequence from clean to polluted is *Fragilaria capucina—Achnanthes minutissima—Cocconeis placentula—Diatoma vulgare—Surirella ovata—Gomphonema parvulum—Synedra ulna—Nitzchia palea*), it has not been widely used and some workers, eg Edwards *et al* (1972) found it unsatisfactory. Hodgkiss & Kan (1978) also used it on streams in Hong Kong but it was not ideal and they used a Tolerant Species index which gave much better results. It is only one stage better than the older Kolkwitz & Marsson (1908) system.

4.9 Generic index

In order to try and make the simplest possible system, data from British rivers is coupled with information from the literature in Table 4 to see how a generic index would work. A note in Descy & Coste (1990) refers to work by Rumeau & Coste (1988) on a generic diatom index.

Table 4 Diatom genera related to gross water quality zones

Clear/Acid	Slightly nutrient rich	Moderately nutrient rich to nutrient rich	Polluted
Eunotia *Tabellaria* *Pinnularia*	*Hannaea*	*Cymbella* *Gomphonema* *Amphora* *Cocconeis*	*Nitzschia* small spp *Navicula* small spp.

In a very crude way these genera give an indication of the water quality—unfortunately there is no single genus which is confined to the most polluted zone. As with the use of genera only in animal-based indices, the use of species will lead to greater reliability and prediction, see Furse *et al* (1984).

4.10 Data from British rivers

Sampling of British rivers has confirmed the overall occurrence of similar groups of indicator species as found in Continental waters (Round 1989). The analysis of the data so far enables a very broad distinction into five zones and some sub-zones. These will be briefly discussed and related to work in other regions. It is possible to quote single species typical of each, but for an overall useful system it is better to give a number of species characteristic of each zone, since few sites have a flora with absolute dominance by a single species. These broad zones are based on downstream surveys of approximately 67 rivers and should be considered merely as preliminary 'broad brush' pictures. Obviously many subtleties will exist above and below point sources of inputs.

Rivers form a chemical and biological continuum, Cushing *et al* (1983) and whilst each species has a range of occurrence in the continuum the ranges of individual species and groups of species will overlap (see the example Fig 8 from Watanabe *et al* (1988). To assume otherwise would be unrealistic though it is still possible to be positive about many distribution patterns—in spite of Cushing *et al*'s comments that 'separate discrete reaches are of little ecological value'—a view which is not supported by the more detailed studies such as that of Leclercq (1987). As Keithan *et al* (1988) wrote 'Given a

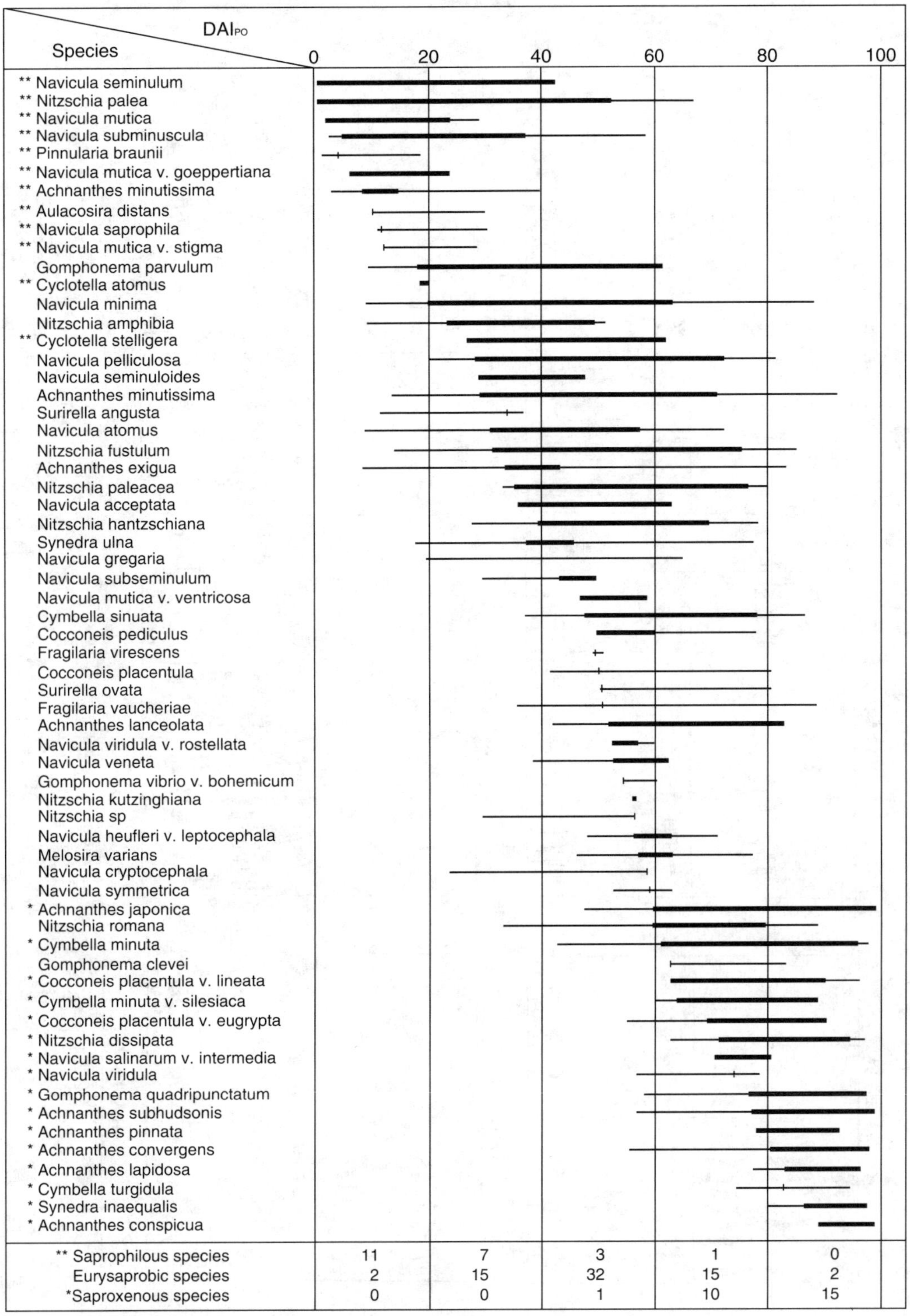

Pollution spectrum by the rise and fall of dominant species accompanying by the variation of DAIPO values

▬▬▬ : range of DAIPO values where species occurred as dominant

——— : range of DAIPO values where species occurred with over 10% relative abundance

Fig 8. The distribution of diatoms across the range from polysaprobic (0–20) to saproxenous (80–100), from Watanabe *et al* (1988).

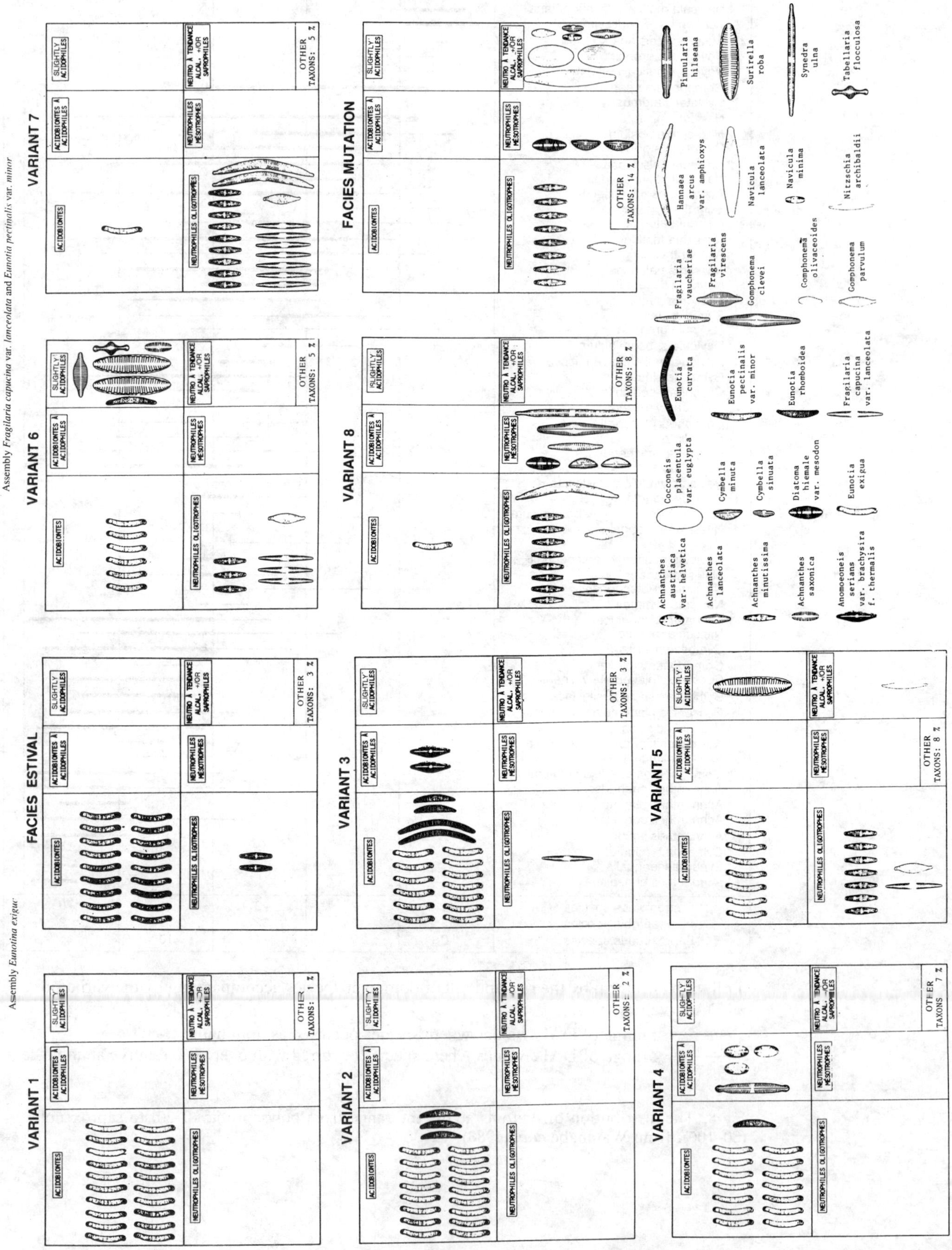

Fig 9. The range of diatom associations in some acid clean waters in Belgium (from Leclercq). Each valve in the diagrams represents an annual value of 5%.

river continuum with longitudinal differences in physical and chemical parameters (current, light, pH, nutrients) qualitative differences in the distribution of algal populations should be expected'—one might even say that this would be inevitable. French experience, Verneaux (1981), using the distribution of animals results in 10 typological levels down a theoretical flowing water system from source to mouth, and the very extensive statistical survey of British rivers, Furse *et al* (1984) produced sixteen similar subclasses.

4.11 Suggested subdivision of river zones based on data from British rivers and related to work in other regions

Zone 1 *Clean water in uppermost reaches (low pH).*

Dominants: *Eunotia exigua* and *Achnanthes microcephala*. These are both very small-celled species, attached firmly to the stone surface. In this zone there is little development of an associated mucilage community, perhaps because flow rate is usually high. Leclercq (1987) provides a detailed analysis of the diatoms in this zone (see Fig 9) ranging from pure dominance of *E exigua* in waters of pH 3.6–4.1 and no alkalinity through five more variants in which species such as *A minutissima*; (some workers do not distinguish *A microcephala* which is present in British rivers from *A minutissima*—this is one of the taxonomic problems yet to be solved); *Eunotia rhomboidea; E curvata; Achnanthes austriaca; Anomoeoneis serians;* and *Pinnularia hilseana* occur. Samples identified by Leclercq also show clearly the appearance of some species identified in Zone 2 of the suggested scheme in small numbers in the 5th variant—ie with pH reaching 4.9–5.2. Although pH is quoted in many works, the values, are only really valuable at the extremes, eg *Eunotia exigua* at pH 3.5 in mine waste polluted streams, Besch *et al* (1972) and down to pH 2.2 in the waters investigated over a wide geographical range by Whitton & Diaz (1981). Correlation with alkalinity and conductivity is much more valuable. It is interesting that none of the species in this group occur amongst the data reported by Watanabe *et al* (1988) for saproxenous taxa. A detailed survey of streams of this type has been made, Round (1991) and as in Leclercq's work, subgroups can be clearly distinguished.

Zone 2 *Nutrient richer and somewhat higher pH*

Dominants: *Hannaea* (=*Ceratoneis*) *arcus, Fragilaria capucina* (var. *lanceolata*) and *Achnanthes minutissima*. Species which may also be common in this zone are *Tabellaria flocculosa* and *Peronia fibula*. Turning again to Leclercq (1977), he has exactly these species in his sub-groups where pH is ranging between 5.6–7.1 and alkalinity 2.8–5.7 (expressed as mg/l $CaCO_3$). He also found sites in which species such as *Cymbella minuta, Diatoma mesodon* were appearing in the flora. Sabater & Sabater (1988) working on rivers in NE Spain also have an upper clean water flora of *Diatoma mesodon* (=*D hiemale* v *mesodon*), *Hannaea arcus* (with *A minutissima*), *Cymbella minuta* (=*C ventricosa*) and *Gomphonema angustatum* v *producta* also present); Cazaubon & Loudiki (1986) working in clear Corsican waters have *Achnanthes minutissima, Hannaea arcus* and *Cymbella minuta* as important diatoms; Backhaus (1968) has *Diatoma mesodon* followed by *Hannaea arcus* in the uppermost reaches of the Danube, Central Europe. These waters do not have the *Eunotia/Tabellaria/Peronia* component, which is not surprising since their bedrock is calcareous. Dell'Uomo (1981) working on the flora in the Appennines (Italy) reports *Hannaea* and *Diatoma* in this region. Several workers also record *Meridion circulare* here and indeed one of the earliest works, Symoens (1957) on Belgian rivers has an association of *Diatoma mesodon/Meridion circulare* in such waters. In some Scottish rivers a few of these zone 2 species (eg *H arcus*) are present along the river indicating good quality water throughout. Sumita & Watanabe (1983) also report *Diatoma hiemale/Ceratoneis arcus* in the upper stations of a Japanese river which has pH values above 7.0. *Diatoma hiemale* and *Meridion circulare* can occur in highly calcareous springs, see Hornung (1959) but they disappear whenever pollution occurs.

Zone 3 *Nutrient rich (corresponding to pH 6.5–7.3, alkalinity 5.0–23.3 in Leclercq (1977, 1988)*

Dominants: *Achnanthes minutissima* is the dominant in the upper region followed downstream by *Cymbella minuta* in the mid-region and in the lower reaches *Cocconeis* *placentula, Reimeria* (= *Cymbella*) *sinuata* or *Amphora pediculus* with a tendency for these to occur in sequence downstream. All these species occur in the eurysaprobic cluster in Watanabe *et al* (1988). This sub-zonation is fairly complex and requires much more detailed analysis. In eutrophic rivers such as are found in most areas of Britain, these subzones occupy almost the whole length of the river. All the species are found widely in European rivers—eg the higher pH sites distinguished by Leclercq are dominated by *A minutissima* and all these species (except *C minuta*) occur in Lange-Bertalot's (1979) Groups 3a/b (he does not deal with waters of the above Zone 1 and 2 character). Many of the listings of species in Lange-Bertalot and other publications include a large number of epipelic species in this zone and although they are found on and in the silt on stones, they are not strictly epilithic.

Zone 4 *Eutrophic with restricted flora due to detrimental influx of materials*

Dominants: The dominant indicator for the decline in water quality is *Gomphonema parvulum*. This zone is also usually distinguished by the relative absence of the *Amphora/Cocconeis/Reimeria* set of taxa. Too few sites of this kind (and of Zone 5) have been found and a more intensive effort is required to improve the typification—unfortunately the sites are often canalised and difficult to sample.

In a paper dealing with the diatoms in an outflow from a sewage treatment plant, Scharf (1984) observed that diatoms were more subtle indicators of the water quality than chemical data and he records some of the Zone 4 (and 5) indicators; these reflect the level of nitrogen compounds, eg at low nitrogen levels *Navicula mutica* was common and as the nitrogen increased the flora changed to *N seminulum/minima—Gomphonema parvulum—Navicula accomoda/N atomus—Nitzschia palea/N thermalis*. Evans & Marcan (1976) experimented with stones placed in troughs through which river water flowed—this resulted in a flora with *Diatoma vulgare, Achnanthes minutissima* and *Cymbella ventricosa*, ie corresponding roughly with Zone 3 outlined above, but when sewage effluent was added in varying amounts up to raw effluent, these species decreased and *Nitzschia palea, Gomphonema parvulum*, small *Navicula* spp and small *Nitzschia* spp increased, ie giving a Zone 4 effect.

Zone 5 *Flora grossly restricted by detrimental influx of materials*

Dominants: The flora is usually dominated by small species of *Navicula* (*N atomus, N pelliculosa*, etc) and small *Nitzschia* spp (*N palea* is universally quoted as dominant but several others are present). Definitive identification of these minute species is difficult and not necessary—their presence is sufficient. However, they are usually accompanied by *Gomphonema parvulum* (except in the most affected sites), *Navicula goeppertiana, Gomphonema augur, Amphora veneta*, and *Navicula accomoda* (the latter dominant in highly affected continental sites but only found in one river in Britain so far). This zone differs from the others in that the characterisation (leaving out the small *Navicula/Nitzschia* group) depends on the occurrence of *N goeppertiana, Gomphonema augur* and *N accomoda* which have to be enumerated, at medium magnification (×40 objective)—since these species rarely achieve any degree of overall dominance in the flora they would rarely figure in counts done under a ×100 objective. All these species (except *G augur*) appear in Lange-Bertalot's (1979) Group 1 and in Watanabe *et al* (1988) indicating extremely poor water quality (saproxenous in the Japanese work).

It is possible to extract from the data in the literature a group of species which would characterise a 6th zone determined by salinity, eg from salt mining/manufacturing perturbation. This would include *Synedra pulchella, S tabulata, Achnanthes brevipes, Amphora coffeaeformis*. See also Kadlubowska (1985) and Jahn & Wendker (1987). In the lowermost reaches these (and other species) indicate the extent of saline water incursion.

In a paper summarising physical/chemical factors on diatom distribution, Evenson *et al* (1981) recorded *Achnanthes minutissima. Cymbella minuta* (var *latens*) and *Fragilaria capucina* all occurring in the lower quartile of measured chemical factors and *Gomphonema parvulum, Navicula cryptocephala* (var *veneta*), *N tripunctata, Nitzschia dissipata, N frustulum* in the upper quartile— the *Navicula* and *Nitzschia* are all part of the silt flora and their distribution in British rivers agrees with the American data; there is thus a generally comparable reaction to eutrophication on a world-wide basis. Work in other regions, eg in central Europe, has shown much the same sequence of species as in British rivers, eg Kawecka (1981) working on mountain streams found *Diatoma hiemale* (and var *mesodon*), *Tabellaria flocculosa, Achnanthes microcephala, A minutissima* in upper reaches, these with the addition of *Hannaea arcus* and *Cocconeis placentula* in the lower alpine reach and *Cymbella minuta, C sinuata* and *Gomphonema intricatum* in the lowermost reach.

According to Lange-Bertalot's study, his Group I species (see Table 5) characteristic of polysaprobic sites cannot be used as 'differentiating species' since they occur ('although not predominant') outside these sites. However, the critical phrase here is 'not predominant' and if percentage occurrence (ie based on counts) is used, the majority of these species will be found to be good indicators of organic (and possibly other types) of pollution. His statement 'Their occurrence alone cannot indicate a distinct level of saprobity; is true but their occurrence in quantity certainly does. Turning to Lange-Bertalot's α-mesosaprobic group, he states that these can act as 'differentiating species' since they 'exist with high reproduction rates' in α-mesosaprobic but *not* in poly-saprobic. So here he is using absence from polysaprobic as a criterion—few workers quote absence as a characteristic, but there is absolutely no reason for this and, in fact, in a hidden manner it is used in indices in general. His IIa group inclines to the waters with higher saprobity than his IIb group. The groups IIa and b are much more widely spread in rivers in general and are not categorised by Lange-Bertalot (see Table 5).

Table 5 has been compiled by extracting from Lange-Bertalot (1979) the species most commonly found in epilithic habitats in Britain, together with some of the indicators used by Watanabe *et al* (1988) and the zonation derived from British rivers.

Many other species especially those in Group IIb are epipelic and not true epilithon—they cannot therefore be used since their % occurrence will be low.

Table 5 Comparison of diatom flora recorded by three different workers using differing systems to group the water characteristics

Lange-Bertalot	Watanabe	British rivers
Group I	**Saprophilic**	**Zone 5**
Most tolerant of pollution		
Amphora veneta	*Navicula goeppertiana*	*Amphora veneta*
Gomphonema parvulum	*Achnanthes minutissima*	*Gomphonema augur*
Navicula accomoda	var. *saprophila*	*Navicula accomoda*
N. atomus	*Navicula minima*	*N. goeppertiana*
N. goeppertiana	*N. mutica*	*Nav.* small spp
N. saprophila	*N. seminula*	*Nit.* small spp
N. seminulum	*Nitzschia palea*	
Nitzschia communis		
Nitzschia palea		
Synedra ulna		
Navicula minima		
N. frugalis		
N. permitis		
N. twymaniana		
N. umbonata		
N. thermalis		

Table 5, continued

Lange-Bertalot	Watanabe	British rivers
Group IIa	**Eurysaprobic**	**Zone 4**
α-mesosaprobic	*Achnanthes lanceolata*	*Amphora pediculus*
Achnanthes lanceolata	*A. minutissima*	*Gomph. parvulum*
Cymbella ventricosa	*Cocconeis placentula*	*Reimeria sinuata*
Diatoma elongatum	*Gomphonema parvulum*	*= Cymbella sinuata*
Fragilaria vaucheriae	*Navicula atomus*	
Navicula avenacea	*Nitzschia frusulum*	
N. gregaria		
N. halophila		
Nitzschia amphibia		
Surirella ovalis		
Synedra pulchella		
Group IIb		
Less tolerant		
Gomphonema pseudoaugur		
Navicula cincta		
Group IIIa		**Zone 3**
β-saprobic		*Ach. minutissima*
Achnanthes minutissima		*Cocc. placentula*
Amphora pediculus		*Cymb. minuta*
Cocconeis placentula		
Diatoma vulgare		
Gomphonema olivacea		
Group IIIb		**Zone 2**
Cymbella sinuata		*Ceratoneis arcus*
Denticula tenuis		*Diatoma mesodon*
Fragilaria capucina		*Merid. circulare*
Gomphonema angustatum		
Navicula mutica s. str.		
	Saproxenic	**Zone 1**
	Achnanthes conspicua	*Ach. austriaca*
	A. convergens	*Eunotia exigua*
	A. japonica	*Pinn. hilseana*
	A. subhudsonensis	*Tab. flocculosa*
	Cymbella minuta	
	Nitzschia dissipata	

The data used by Lange-Bertalot from the wide, silted rivers unfortunately contain a large number of epipelic species and the ease of use of the system is due to the fact that it is really only a statement of what does and does not occur in the polysaprobic zone. The vast masses of waters fall outside this and therefore 'differentiating species' for the degrees of 'pollution/saprobity/eutrophication of rivers' need to be extracted from Groups II and III; it may be that the diatoms in regions so characterised need further detailed subdivision. Lange-Bertalot considers that the species groupings react to organic or chemical pollution in a roughly similar manner. The diagram (Fig 8) reproduced from Watanabe *et al* (1988) is considered the best example known illustrating the gradation of species which can be expected as sampling proceeds from waters of poor quality (saprophilous) to pristine (saproxenous). The heavy bars would be the distributions if counting was limited to 100 cells and only the dominant (indicator species) were considered. Not all of his species occur in British rivers and a finer subdivision of the clean waters can be achieved, cf. Lerclercq (1988). The distribution of all eurysaprobic species need to be carefully compared with chemical conditions and this may lead to a better defined system.

4.12 General discussion of Indices

A thorough study of the published data and experience of sampling British rivers reveals that in spite of some comments to the contrary all the studies are essentially based on nothing more than 'indicator' species. The presence or absence of these provides very powerful indications of water quality expecially when combined with cell count data. Mere recording by + or − in tables is only useful for the occurrence or otherwise of the indicators of extreme conditions, ie the very pure or very polluted waters. In ecological surveillance there will always be odd counts (stations) which appear to upset the general pattern—these may be due either to contamination (at the site or during preparation of material in the laboratory) or due to unusual and unexpected conditions at a station—only if this 'anomaly' is persistent, do the causes need to be investigated. In much of the literature the 'indicator' species have to be searched out of masses of scattered records of minor species—in almost every case they are present, and if sampling had been of material from well defined microhabitats and counting had been reduced, the pattern would have been clear—a simple example of confusion is when tables are constructed using both epilithic and epipelic data. The plant sociological approach (which itself is based on indicator species) will reveal the real subtleties of epilithic diatom distribution. It has been achieved at the clean water end, see Leclercq & Manquet (1987) but it is essential that finer details of distribution in the mid/lower river reaches is achieved. This is quite obvious from Table 5 where zones in British rivers were derived purely from downstream studies, ie starting the sampling as near to the source (water as pure as possible for the stream!) and proceeding downstream to the estuarine reach (assuming this to be the richest (most eutrophic) site (zone 5)—in fact, the data showed that the polluted zone 5 often extends upstream into the mid-reaches. The position of the diatoms in these zones is, in fact, in general agreement with those of the other workers, eg Backhaus (1968) who showed a clear set of changes downstream and overall correlated with the data on pH, alkalinity, calcium concentration and conductivity. The separation of the effects of eutrophication, heavy metal contamination and organic pollution is a major problem though in some rivers these all tend to coincide—see comments in Descy *et al* (1982). The numerical indices reveal that none of my groupings clash with these but the numerical indices (especially those of Descy and of Leclercq & Manquet) do not separate the large mid-region of water quality adequately. Descy & Coste (1990) recognised this feature and concluded that the CEE 88 format would be preferable. It does distinguish the regions of water quality better than the earlier indices but they do also stress that the method depends upon skilled diatomists or at least high quality technicians. The moderate/extreme pollution zones are also well characterised by indicator species.

The construction of the indices is a simplification of the recording of the indicator species into a brief numerical classification, usually 5 classes but sometimes 4, eg Wegl (1983) and up to 10, eg Sladacek (1973). Cook (1976) gives an excellent summary of many methods and concludes that the Chandler Score invertebrate system is fairly good but this type of index has not been tested with diatoms. In a discussion mainly based on invertebrate sampling Cook did not recommend the Lothian or Treat Biotic indices or the indices of diversity. This is acceptable but more attention should be given to seeking out further 'indicator' species. The system could be improved especially in the middle/lower reaches of rivers. The difficulties here are almost certainly associated with levels of nutrients rather than the extreme factors applying at either end of the river continuum.

Many workers have related their indices (1–5) to levels of various chemicals or allied them to saprobity/eutrophication/pH-alkalinity-calcium levels eg Hartmann (1983); see Fig 9, Watanabe *et al* (1988), Leclercq & Manquet (1987), Steinberg & Schiefele

(1988). Descy *et al* (1982) also used calcium carbonate levels to divide up the waters of the R Meuse. Conductivity measurements also seem to be good measures to relate to diatom distributions, see data in Negoro & Gotoh (1983). Schoeman (1976) found a good fit between groups of diatoms and chemical levels in a South African river with species similar to those in Europe. House (1989) has made a detailed analysis of factors used in river water quality measurements (Table 6) to give four chemical indices and using only the Water Quality Index has defined four water quality classes which can be also related to the National Water Council (NWC) classification, showing an excellent fit between the two.

Table 6 The factors used to devise four indices of water quality (from House 1989)

WQI	PWSI	ATI	PSI
Dissolved oxygen	Dissolved oxygen	Dissolved copper	Total copper
Ammoniacal nitrogen	Ammoniacal nitrogen	Total zinc	Total zinc
Biochemical oxygen demand	Biochemical oxygen demand	Dissolved cadmium	Total cadmium
Suspended solids	Suspended solids	Dissolved lead	Total lead
Nitrate	Nitrate	Dissolved chromium	Total chromium
pH	pH	Total arsenic	Total arsenic
Temperature	Temperature	Total mercury	Total mercury
Chloride	Chloride	Total cyanide	Total cyanide
Total coliforms	Total coliforms	Phenols	Phenols
	Sulphate	Total hydrocarbons	Total hydrocarbons
	Fluoride	PAH	PAH
	Colour	Total pesticides	Total pesticides
	Dissolved iron		

WQI = Water quality index
PWSI = Potable water supply index
ATI = Aquatic toxicity index
PSI = Potable sapidity index

More recently Descy & Coste (1990) report on a grouping of French waters based on clustering of chemical features and their conclusion is that the diatom indices fit well with the chemical indices. They distinguish six types of pollution and each is subdivided into five classes. 1. Organic (O_2, NH_4-N, BOD5, DOC, TOC); 2. Eutrophication (PO_4); 3. metal (Cd, Cu, Cr, Hg, Ni, Pb, Zn); 4. Inorganic (fluoride, NO_2-N, NH_3-N); 5. Organic (two sets of measures (a) CN, phenols, hydrocarbons (b) anionic detergents; 6. Salinity (conductivity, SO_4 and Cl). These chemical systems should now be related to the occurrence of diatom indicator species since as shown in the pristine water quality region, especially by Leclercq (1988) much finer distinctions may be possible using the biological data. For example, the very slight drainage from veins of rock containing carbonates was immediately detectable in two streams in a predominantly acid region—the diatoms *Reimeria sinuata* and *Cymbella minuta* acting as the indicators, Round (1991). In the discussion of House (1989), a table reveals great variations in the numerical data in the WQI such that some values in the pristine class can re-appear in the polluted class—due to one-off events prior to sampling. These give important clues to detrimental events but may also distort the interpretation, eg in one river over three years the quality varies over three water quality classes—it would be most instructive to see what effect this has on the diatom indicator species. There is no reason why the data should not simply be used in the form of a sequence of indicator species reading from pure to polluted or the reverse. At the lower end of the pH scale, the diatoms can be correlated with pH ranges, but above pH 5.5 this becomes difficult until very high pH values are reached. It is, however, possible to indicate an upper limit for the acidophilic species and a lower limit for the alkaliphobic. The majority of river sites lie in the mid-pH region and correlation here is difficult. The range of species is better correlated with carbonate hardness, see Hartmann (1983) for river data—unfortunately his data is not from diatoms occupying precise habitats but from material collected from plant/woody material hanging in rivers. The danger of proposing a close relationship with abiotic factors is that pure through polluted can encompass many different states, eg acid pure/alkaline pure through to nutrient rich/pollutant rich/high salinity. Placing values on a continuum could be misleading—the values are most meaningful at the extremes. Nevertheless it is worth attempting a scheme with the *proviso* that it is only a preliminary stage. The system used by Leclercq (1987) is a good example (reproduced here, Fig 9). It would be convenient if indicator diatoms could be found for specific factors, but at the moment the necessary data is rare, Leclercq & Manquet (1987). In the system of Steinberg & Schiefele (1988) the water classes derived from the diatom data showed a reasonable relationship with phosphorus levels. They have two groups of species related to low and high nutrient levels but much more information is required to refine this system especially since they include species from the silt flora on the stones.

There is frequent use of ordination techniques such as Principal Component Analysis, eg by Sabater & Sabater (1988). Most of these seem to have been done simply because it is fashionable, and the results show nothing more than is concluded from the tabulated data. Equally, the techniques to cluster sites or groups of species in dendrograms adds almost nothing to the overall information. Either the techniques are of little value or the workers are not able to extract useful information from the analyses.

In almost all studies it is obvious that more time could be spent doing closer interval sampling and reducing the time spent on cell counting—the major species are the most useful indicators—a rapid scan of the microscope slide could be used to detect any very specific indicators present in small numbers, but that data has to be used with great care. There are several papers criticising the use of diversity indices, eg Hulburt (1971), Murphy (1978), Katoh (1991). The last author concluded that a 'saprobity' type index such as that of Pantle & Buck (1952) was superior to diversity studies. Washington (1984) provides a detailed discussion of mathematical indices and found only three out of eighteen diversity indices of value. Before lengthy quantitative/mathematical techniques are used, a careful reading of Fryer (1987) is recommended.

Some studies resulting in criticisms of the index system, eg Caspers & Schultz (1960, 1962) are flawed since old indicator tables, eg the early Kolkwitz & Marsson (1980) or the Palmer index (1969), are not sufficiently accurate.

5 Future work

The work to date reveals clearly the association of diatom species with (1) clean/nutrient poor waters, (2) with eutrophic waters and (3) with polluted waters. The first group has already been sub-divided by Belgian workers and the data show that very subtle distinctions can be made and related to chemical features of the waters. Further studies of the flora in such waters would lead to early detection and early warning of biological changes. Some waters in this region are sensitive to 'acid rain', and the flora of these requires study coupled with recovery studies based on liming, eg as in some small lakes, Round (1990). Much work is required on the second group of waters which is likely to prove the most difficult to sub-divide. Broad categories of eutrophication need to be linked with diatom indicator species and then the effect of specific pollutant conditions determined. The third class of waters also requires careful analysis of the diatom associations in relation not only to the effects of individual pollutants but also in relation to recovery from pollution, eg we do not know the time it takes for the pollution tolerant species to die out when the pollutant source is removed.

The epilithic diatom flora has been studied by some workers in relation to light levels (shading), current speed, geological nature of substrata, grazing, etc. and all these features still require further investigation, singly and in combination. In addition, the effect of water depth has been relatively neglected and sampling across the channel of some rivers should be undertaken.

The diatom flora of waters arising on acidic, neutral and alkaline (including chalk streams) needs determining since it appears from the literature that some species are tolerant of all conditions. The gradual effect of downstream eutrophication of each class of water needs further study.

The problem of the 'forest' growth on silted stone surfaces needs more attention, in particular, with regard to the removal of the silt flora and characterisation of water quality only by use of the true epilithon.

Finally, a study of epipelon, epiphyton and epipsammon is overdue, since we know little of the relationship between the diatom flora of the communities occupying these niches and water quality. One of other of them may yield valuable data on water quality, especially in the lowland, slow flowing reaches where sampling of stone surfaces is sometimes difficult.

Although there is certainly overlap of species between the four benthic diatom communities, only a detailed study of each will enable a proper assessment. This is needed to provide an overall understanding of distribution of diatoms in rivers and will enable a simple but comprehensive set of indicator species to be established for all types of rivers.

There are many developments which could be included after further detailed studies, eg constructing computer-based programmes linked to a data base, automated counting systems and working to a long-term integrated suite of methods for animals and plants. Biological monitoring is the logical step to aid interpretation of chemical data and as many complementary techniques as possible are of great value when monitoring eutrophication and pollution.

6 Appendices

Appendix 1
Example of recording sheet

Species	Dates or sampling site							
	1	2	3	4	5	6	7	8
Tabellaria flocculosa								
Eunotia exigua								
Diatoma mesodon								
Achnanthes microcephala								
Hannaea arcus								
Fragilaria capucina								
Achnanthes minutissima								
Achnanthes lanceolata								
Cocconeis placentula								
Amphora pediculus								
Synedra ulna								
Cymbella minuta								
Reimeria sinuata								
Navicula gregaria								
Gomphonema parvulum								
Gomphonema augur								
Navicula goeppertiana								
Amphora veneta								
Navicula accomoda								
Small *Navicula* spp								
Nitzschia palea								

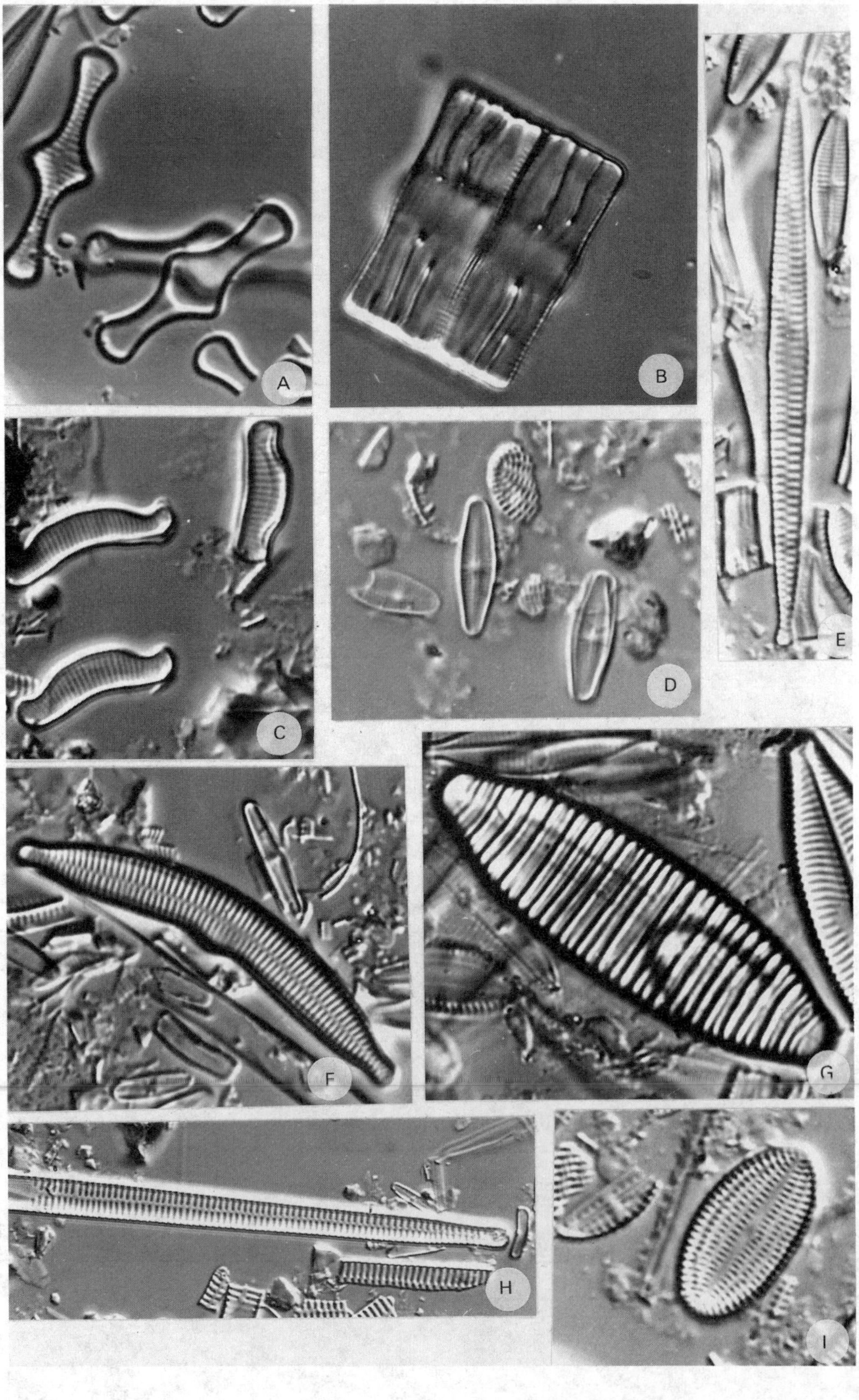

Light micrographs of common diatoms in river epilithon. A: *Tabellaria flocculosa*—valve (left) and girdle bands (right). B: Two cells *Tabellaria flocculosa* in girdle view. D: *Eunotia exigua*. D: *Achnanthes minutissima*. E: *Synedra* cf *capucina*. F: *Hannaea arcus*. G: *Diatoma vulgare*. H: *Synedra ulna*. I: *Cocconeis placentula*.

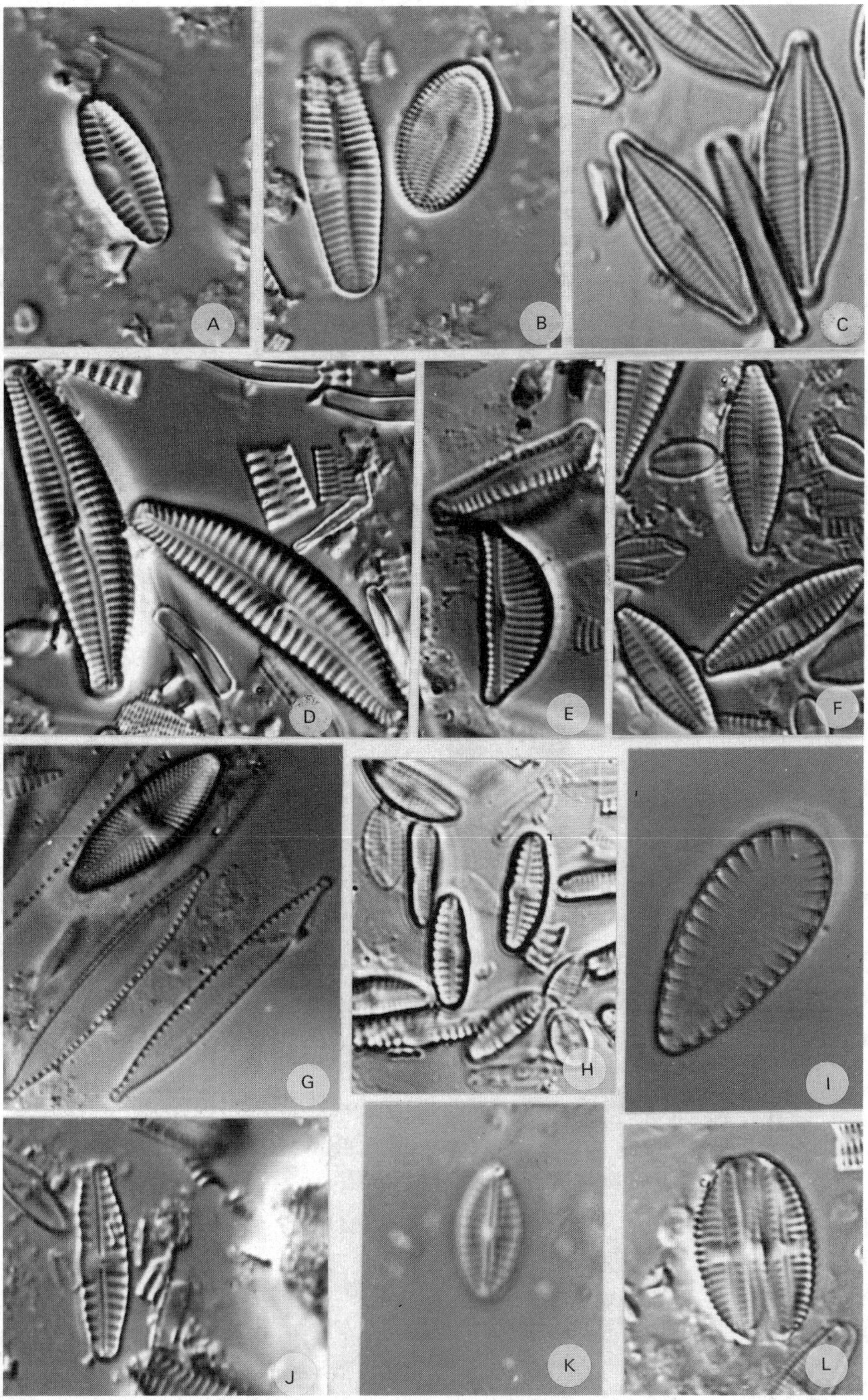

Appendix 2 continued. A: *Achnanthes lanceolata*, B: *Achnanthes lanceolata* and *Cocconeis placentula*. D: *Navicula* sp D: *Cymbella affinis*. E: *Cymbella minuta*. F: *Gomphonema parvulum*. G: *Navicula goeppertiana* (above) and *Nitzschia palea*. H: *Reimeria sinuata*. I: *Surirella ovata*. J: *Reimeria sinuata*. K: small *Navicula* sp. L: *Amphora pediculus*.

Scanning electron micrographs of common epilithic diatoms. A & B: *Eunotia exigua*, C & D: *Diatoma mesodon*, E: *Tabellaria flocculosa*, F: *Hannaea arcus*

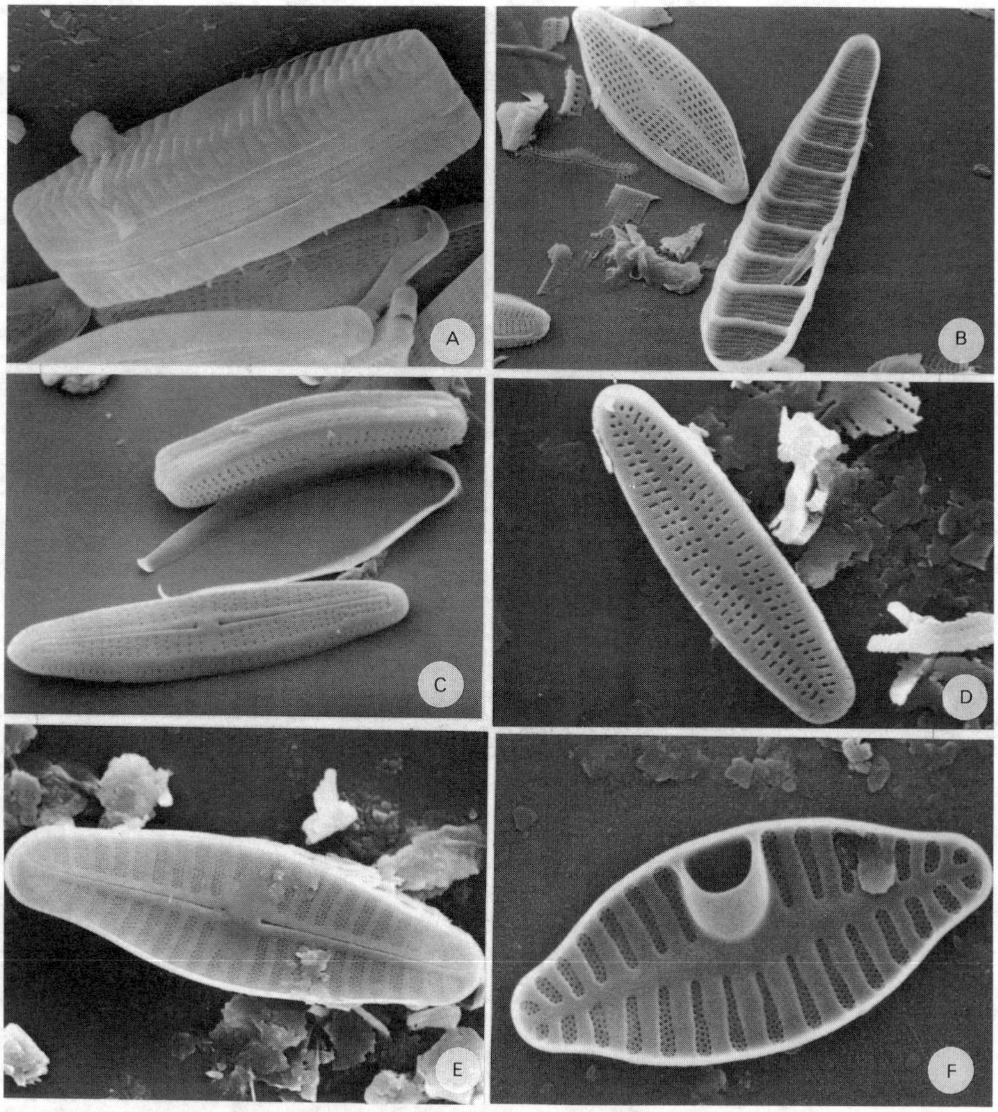

Appendix 3, continued. A: *Fragilaria vaucheriae*, B: *Meridion circulare*, C & D: *Achnanthes minutissima*, E & F: *Achnanthes lanceolata*, outside of raphe valve and inside of non-raphe valve.

Appendix 3, continued. A: *Navicula* sp, B: *Navicula goeppertiana*, C: small *Navicula* sp, D: *Gomphonema parvulum*, E: *Gomphonema augur*, F: *Reimeria sinuata*, G: *Amphora veneta*, H: *Cymbella minuta*.

Appendix 4
Notes on species

Achnanathes microcephala. This is the capitate taxon recognised in the Hustedt floras (1927–56, 1930) but included in *A. minutissima* by Lange-Bertalot & Ruppel (1980). Both this and the succeeding species occur above and below sites contaminated by heavy metal waste from mining, Deniseger, Austin & Lucy (1986). *A. microcephala* occurs at a lower pH than *A. minutissima*, Besch *et al* (1972)—this is confirmed for British rivers.

Achnanthes minutissima. This is one of the most abundant species in streams throughout the world and often a major colonist in experimental systems, (eg Peterson & Stevenson (1989). It is very common in disturbed sites and under high flow regimes. It was eliminated in a stream subject to acid (and metal) mine drainage (pH 3.0–4.1) but occurred on either side of the affected stretch (pH 6–8), Lampkin & Sommerfeld (1982). A study of copper addition in a Californian stream resulted in about a 4-fold increase in *A. minutissima* as the copper eliminated grazers and exposed space as other species were eliminated, Leland & Carter (1984). It is often quoted as the first coloniser of rock and particularly glass slides, Eichenberger & Wuhrmann (1975) and Rushforth & Squires (1986) found it in 96% of the new sites being colonised after the Mt. St. Helens eruption, and it was still abundant six years later, Steinmann & Lamberti (1988). Addition of sewage to R. Ribble water in experimental troughs decreased the abundance of *A. minutissima*, Evans & Marcan (1976). Most workers quote this species as an indicator of clear, well-oxygenated waters and this fits with the present finding at the zone 2/3 boundaries.

The Japanese workers, Mayama & Kobayasi (1984) recognise a variety of *A. minutissima* v. *saprophila* which is abundant only in the most (organically) polluted waters.

Achnanthes lanceolata. This is generally found in nutrient rich waters at higher pH/alkalinity than the other two. It is dominant on stones in many lowland waters, eg in East Holstein, Raabe (1951). Many varieties and forms have been described and further work is needed to relate these to water quality states—some varieties are tolerant of moderate pollution.

Amphora pediculus. This is a complex of forms, see Lee & Round (1989) but essentially all are very small and attached closely to rock surfaces.

Amphora veneta. This very characteristic species only occurs in Zone 5 but seems to be rather rare in British rivers.

Brachysira serians. This distinctive taxon is not abundant in British rivers, but is characteristic of slow-flowing acid waters draining bog and moorland.

Cocconeis placentula. Very common on artificial surfaces and on stones in mid-reaches of rivers but seems to be eliminated in high nutrient zones. Pringle & Bowes (1984) found that experimental blocks with and without added nitrate supported rich *C. placentula* growths, but it was not abundant when PO_4 was added. There are many morphological variants of this species—they require further study to determine any particular preferences.

Cocconeis pediculus. This is a dominant species epiphytic on *Cladophora* and is hardly ever found on stones—though recorded as such in some works. When it is recorded in the literature there is often the mention of *Cladophora*, eg in Cazaubon (1988)

Diatoma mesodon (= *D. hiemele* var. *mesodon*). In clean water, fast flow. Usually in acid regions but also requiring slight $CaCO_3$ content. Also reported in some clean higher pH rivers. This taxon is often abundant in springs where $CaCO_3$ concentrations are high, Hornung (1959). So although often quoted as an indicator of clean, acid waters, it is probably the unpolluted nature and not the pH of the water which is significant.

Diatoma vulgare. Common in eutrophic/polluted reaches. Evans & Marcan (1976) found it in the R. Ribble and it continued to grow well on the addition of sewage effluent to experimental troughs. One piece of experimental evidence not fitting this pattern is that of Evenson *et al* (1981) who found this species correlating with the lower quartile of their chemical factors—but other evidence is against this. Salden (1978) gives a pH range of 7.4–8.0.

Eunotia exigua. This has one of the most distinctive distributions, always being associated with clear, clean low pH waters (eg 3.4–5.6 and much less abundant at 6.0–7.0)—Besch *et al* (1972), Fabri & Leclercq (1979). Reported to be very resistant to Zn and Cu.

Other *Eunotia* spp. (eg *E. tenella, E. lunaris, E. veneris, E. rhomboidea, E. incisa, E. vanheurckii*) all have distributions in acid waters, Patrick & Roberts (1979) and it is not too much of a generalisation to say that their absence from a reach indicates neutral to high pH and elevated alkalinity/nutrient status. One other striking feature is that, coupled with the low pH preference, some are tolerant of high heavy metal concentrations, (eg *E. tenella* became very abundant at mining sites in N. America—Lampkin & Sommerfeld (1982). In a study of colonisation on sandstone and chalk, Lay & Ward (1987) found *Eunotia* spp. only on sandstone—the pH was 6.8 on the sandstone and 7.2 on the chalk, but the $CaCO_3$ values were quite different (4.4 mgl and 103 mg/l respectively) showing again that the carbonate/alkalinity measurements separate associations much more clearly than pH in the neutral range.

Fragilaria capucina. This small needle-like diatom appears just after the *Eunotia exigua* zone. There are several forms. Hürlimann & Straub (1991) recognise 7 variants and these do have important relationships to nutrient status.

Fragilaria vaucheriae. A very common species in the eutrophic zone of rivers. Several varieties and forms exist, each probably with specific requirements which need study.

Gomphonema angustatum. This species is often confused with *G. parvulum*, but is generally larger and with less striae in 10 μm than found in *G. parvulum*; the central isolated punctum is also more obvious in *G. angustata*. Not usually extending into organically richer, polluted waters and usually more abundant in alkaline situations.

Gomphonema olivaceum. Several varieties and even species have been united in this taxon, see Krammer & Lange-Bertalot (1986). A form with four distinct puncta around the central raphe endings is very common and has been separated by some workers as *G. olivaceum* v. *minutissima* (syn. *G. olivacioides* or *G. tetrastigmatum* or *G. quadripunctata*).

Gomphonema parvulum. This is a rather variable species, see illustrations in Krammer & Lange-Bertalot (1986) and as a result is often quoted as having a rather widespread distribution. However it often occurs in quantity in 'polysaprobic' waters along with other indicator species (coupled with the lack of the less tolerant-to-pollution species). Forms occurring outside such waters will be accompanied by species characteristic of moderately eutrophic waters. It did increase with addition of sewage in Evans & Marcan's (1976) study, but not as strongly as some other species. This would seem appropriate from experience of British rivers where it is common in moderate pollution but less abundant in extreme pollution.

Hannaea (= *Ceratoneis*) *arcus*. A species of clean river water. It is one of the most distinctive diatom taxa, and is unlikely to be confused with any other. Several forms have been described, but little is known of their ecological preferences or indeed their validity. Also abundant in calcareous rivers, eg in mountain streams of the Pyrenees, Sabater & Sabater (1988), Sabater *et al* (1987) and Cambra (1987) at pH 7.2–8.6 and conductivity 100–300 μS cm^{-1}. Has been used as a type diatom for the Ceratoneito-Hydruretum rivulare in high altitude fast-flowing waters. *Hannaea arcus* was abundant in a study of the diatoms in a river receiving discharge from a reservoir, immediately below the discharge in a zone which was stimulated greatly (growth was four times as great here as downstream)—the one striking chemical characteristic at the discharge site was the high value for NH_4.

Meridion circulare. Another unmistakable taxon especially when the colony form is observed. It tends to occur in similar waters to *Hannaea*, but particularly favours springs, Olive & Price (1977) and calcareous waters. Whilst occurring to some extent on stone surfaces it is more abundant amongst mosses. It dies out as nutrient status increases, but increase in particulate matter may also be detrimental. Often quoted as abundant in rivers, but not found frequently in British rivers.

Navicula accomoda. Rare in British rivers and confined to 'polysaprobic' zones. Characterised by the parallel striae, elliptical valves with slightly capitate apices. Somewhat similar taxa which are found in less polluted waters need checking against *N. halophila* or *N. gregaria*.

Navicula atomus, N. saprophila. Two small forms both recorded commonly in highly eutrophic, organically polluted waters.

Navicula goeppertiana. A very distinctive taxon recognisable from the distinctive large stigma on one side of the central area. A series of forms and other taxa, eg *N. mutica, N. muticopsis, N. cohni* occur in the same sub-group of *Navicula* and may be quoted by some workers since *N. goeppertiana* has been split off from *N. mutica*.

Navicula gregaria. A very confused taxon probably most common in eutrophic waters. This and *N. phyllipta, N. veneta* and *N. cryptocephala* are all confusing but often recorded by workers. They really belong to the sediment flora, but are frequently contaminant on stone surfaces.

Navicula lanceolata and *N. tripunctata* are characteristically large species, again very common in eutrophic situations and really part of the sediment flora.

Navicula subtillisima. A fairly distinctive form but only common in acidic waters draining upland regions.

Nitzschia amphibia and *N. frustulum*. These are small species with distinct transverse striation—the former more coarse (13–18 striae in 10 μm) than the latter (19–30 in 10 μm)—the striae can usually be seen to be composed of pores in *N. amphibia*.

Nitzschia palea. All the evidence points to a distribution in polluted waters. It has also been shown to grow best at high pH (8.2) by Eggleston *et al* (1987), but other workers have suggested pH indifferent, best around 7.0—Lowe (1974) and others have suggested best above 7.0. There are surprisingly few other experimental studies, but in Evans & Marcan's (1976) study it increased markedly in the troughs with added sewage (as did *Nitzschia communis, Nitzschia stagnorum* and *Navicula atomus*). The increased representation of *Nitzschia* (and *Navicula*) may be due to the higher phosphate and to a lesser extent nitrate values in the eutrophic regions—several papers using experimental techniques showed that both biomass and species of *Nitzschia/Navicula* increased when fertilised with phosphorus and nitrogen sources, eg Pringle & Bowers (1984), Keithon *et al* (1988). In N. Spain at polluted sites on the River Ter (with *N. ganderheimensis*) it contributed 80% of total flora, Sabater & Sabater (1988).

Peronia erinacea. A distinctive taxon in relatively pure and often in acidic waters. Easily recognisable owing to the presence of short raphe slits on one valve and the lack of slits on the other. This genus is called *Peroniopsis* in Runeau & Coste (1988).

Reimeria sinuata. Previously placed in *Cymbella*. Easily recognisable by the valve shape and the presence of a stigma in the centre of the valve between the central raphe endings.

Rhiocosphenia abbreviata (= *R. curvata*). This is frequently quoted as a component of the epilithon but like *Cocconeis pediculus* it is more abundant epiphytic on *Cladophora* (a similar correspondence is obvious in Cazaubon's (1988) study).

Surirella ovalis. This comprises a cluster of forms, see Krammer & Lange Bertalot (1987), some of which have been upgraded to species. They are readily recognisable by the very fine striation of the valve surfaces. Their centre of distribution tends strongly to the most eutrophic/organically polluted zones and also extending into brackish water.

Synedra ulna. The only large araphid genus occurring to any extent in the epilithon. Most abundant in nutrient rich zones and virtually absent from the extremes of water quality.

Tabellaria spp. These are only found on stones in acid reaches often where these cross moorland. *T. flocculosa* is the common species but *T. fenestrata* is also recorded in the literature. The genus is rare in the river epilithon as a whole. It certainly does not tolerate any degree of pollution, Hornung (1959).

N.B. These notes are intended to complement the illustrations and are **not** comprehensive in a taxonomic sense. For detailed comment on the latter point, see Krammer & Lange-Bertalot (1986, 1988, 1991a, 1991b). The comments on distribution are based on experience in British rivers and do not always agree with comments elsewhere in the literature.

Appendix 5

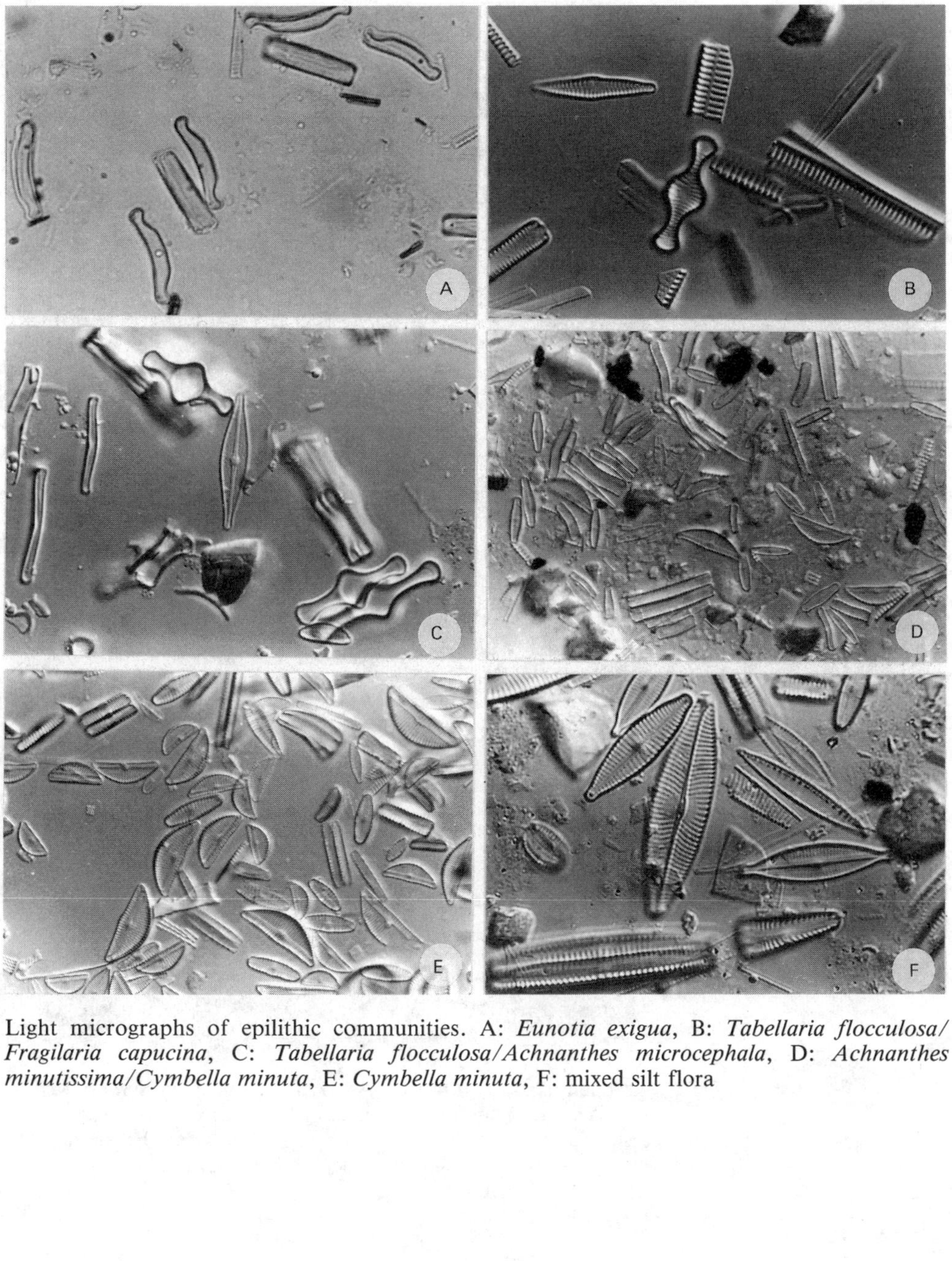

Light micrographs of epilithic communities. A: *Eunotia exigua*, B: *Tabellaria flocculosa/Fragilaria capucina*, C: *Tabellaria flocculosa/Achnanthes microcephala*, D: *Achnanthes minutissima/Cymbella minuta*, E: *Cymbella minuta*, F: mixed silt flora

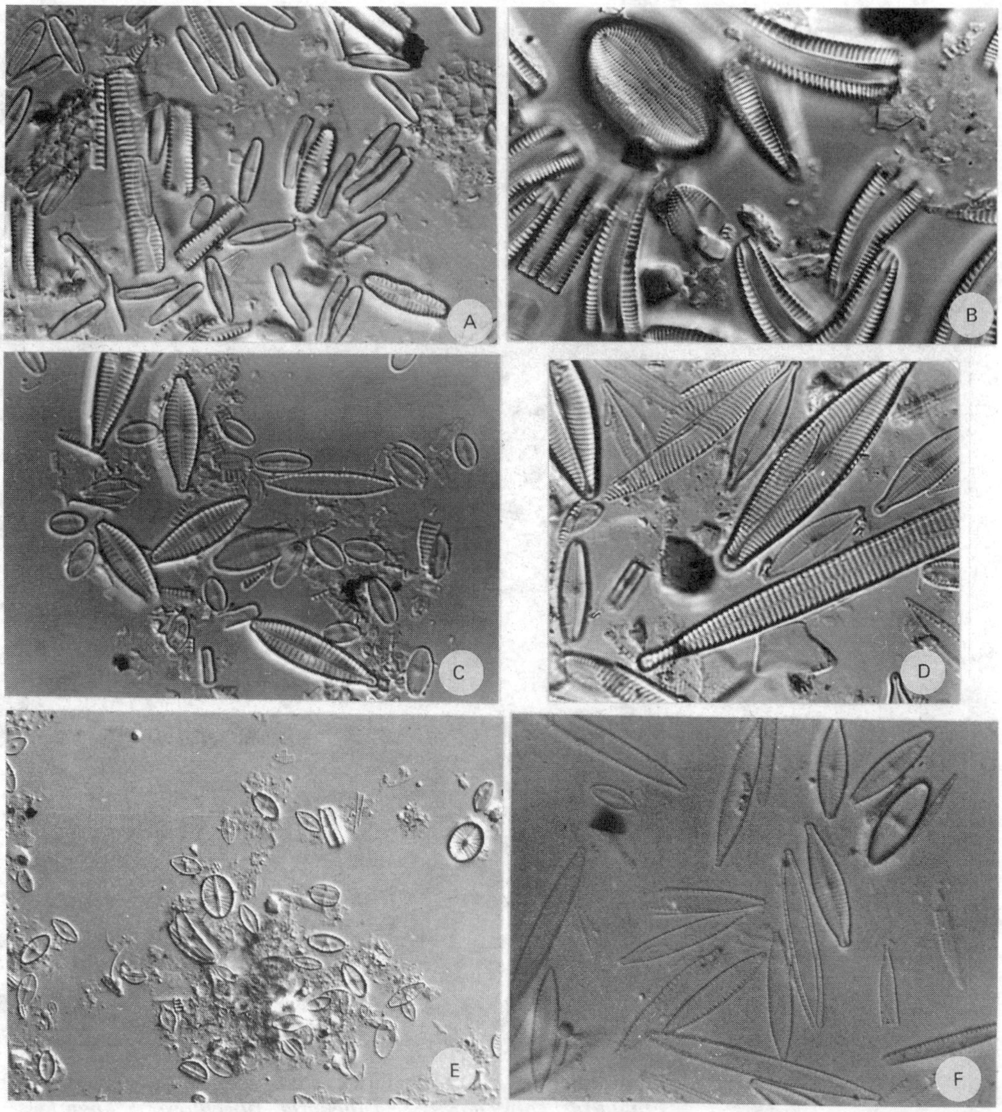

Appendix 5, continued. A: *Achnanthes minutissima/Reimeria sinuata*, B: *Rhoicosphenia/Cocconeis* community off *Cladophora*, C: Gomphonema parvulum/small *Navicula* species, D: silt community, *Navicula* spp./*Synedra ulna*, E: small *Navicula* spp., F: *Nitzschia palea/Gomphonema parvulum/Navicula goeppertiana*.

Appendix 6

Selected epilithic species with saprobic valency (SV) and indicator value (IV), from Leclercq & Manquet (1987). Only the species found commonly in the epilithon of British rivers are listed: NB: This system uses 1 to indicate tolerance of organic/pollution and 5 for the cleanest in the SV column, ie the reverse of the system described in this booklet.

	SV	IV
Achnanthes lanceolata	3.5	1
A. microcephala[1]	5.0	4
A. minutissima	4.0	1
Amphora pediculus	4.0	1
A. veneta[1]	1.0	3
Cocconeis placentula	4.0	2
Cymbella minuta	3.0	1
Diatoma mesodon	4.5	3
Diatoma vulgare	3.0	1
Eunotia exigua[2]	4.5	4
Fragilaria capucina	4.5	3
Fragilaria vaucheriae	3.5	2
Gomphonema angustatum v. *producta*	3.0	3
G. olivaceum	3.0	2
G. parvulum	2.0	1
Hannaea arcus[1]	4.0	2
Meridion circulare	4.0	3
Navicula accomoda	1.0	4
N. atomus	2.5	4
N. cincta †	2.5	3
N. cryptocephala †	3.0	2
N. goeppertiana	1.5	4
N. gregaria	2.5	4
N. lanceolata †	3.0	3
N. phyllepta †	4.0	2
N. saprophila	2.0	2
N. subtillissima	2.0	1
N. tripunctata	4.0	2
Nitzschia amphibia	2.5	2
N. frustulum	3.0	3
N. palea	2.0	2
Peronia fibula	4.5	3
Reimeria sinuata	4.0	2
*Surirella ovalis** †	3.0	3
Synedra ulna	3.0	2
Tabellaria flocculosa	4.0	2

[1] *Not in Leclercq & Manquet (1987), value taken from Descy & Coste (1987) and modified slightly.*

[2] *Other Eunotia* species classified similarly. Also any *Pinnularia* species. These two genera are vitually confined to the upper clear waters.

† These species are not attached forms but are so common in the metaphyton that they are included.

* A mean figure from three varieties in Leclercq & Manquet (1987).

Appendix 7
Defined area sampling

Removal from a defined area has been achieved in a variety of ways. Margalef (1949) devised a system for laying a film of plastic on the stone surface and stripping this and the underlying diatoms off—it has never been used to any extent, but there is potential for development. Goldsborough (1989) used a layer of adhesive tape applied to artificial substrata (acrylic rods) and obtained excellent material suitable for light and especially scanning electron microscopy. Its use on encrusted natural substrata may be less effective. All other techniques (qualitative or quantitative) require isolating an area and then brushing the cells off. For more precise measurements, short lengths of glass tubing can be cut, ground smooth and attached to the dry stone surface with an adhesive and after drying the area inside, the area inside the glass tube can be brushed and the released diatoms washed/pipetted out into centrifuge tubes. Problems arise when attaching the tubes on rough stones and numerous tubes have to be attached to obtain sufficient samples to eliminate the effects of patchiness. All these techniques result in problems if an attempt is to be made to check on live cells, division stages, grazing, etc. The final samples, in distilled water, can be made up to standard volumes and counted by the usual techniques (inverted microscope, Sedgwick Rafter cells, etc) and the results expressed as cell numbers per mm^2 of stone surface. They can also be used for making permanent slides for identification and further counting if difficulties arise with the samples sedimented in water.

Such highly quantitative methods are unnecessary for much basic work, since conclusions from presence/absence and abundance of indicator organisms is all that is required. If, on the other hand, continuous records are required and the balance between species or annual trends in absolute species number are sought, then the quantitative methods should be used. For most water quality assessment the need is for rapid determination and if the floristic differences are so subtle that statistical techniques are necessary to distinguish sites, it is highly unlikely that the sites differ greatly in water quality. It should be remembered that no two sites will ever be exactly the same and any site will have changing proportions of species on a day-to-day or week-to-week basis. Apart from any other habitat based disturbance, the growth rates of individual micro-organisms vary, the timing of growth pulses vary and species which have such short cell cycles (as short as 24h) can produce enormous populations over short periods of time. For these and many other reasons the presence (abundance) or absence of indicator species is the best indication of the state of the water—numerical indices are, in fact, only slightly more sophisticated expressions of presence or absence of such species. There is a pool of species available in all rivers and transport between rivers is common. The species growing under any one set of conditions will be selected from the pool and because of the vast range of conditions available, even in apparently similar stretches of different rivers, it is necessary to list several species for each water quality—any of those occurring as dominants will be sufficient to indicate the water quality. More subtle balances of species will of course provide more increasingly detailed information.

7 Glossary

Acidophilic	Acid loving
Alkaliphobic	Having an aversion to or avoidance of an alkaline environment
Angiosperm	A flowering plant.
Artificial substrata	Any object (glass, plastic, wood, etc) placed in a river for experimental purposes, eg on which to grow details.
Autotrophy	The ability of organisms to synthesize their organic requirements from inorganic precursors using carbon dioxide and nitrates of ammonium salts as sources of carbon and nitrogen.
Auxospore	The large cell produced after sexual fusion in diatoms.
Benthos	The populations of organisms living associated with underwater surfaces. Strictly living on the bottom.
Biomass	The total amount of living material (of species or groups of species) in a habitat.
Bryophyte	A moss or liverwort.
Calcareous	Containing calcium carbonate; chalky.
Capitate	Terminating in a head or knob.
Cell cycle	The process of cell production from one reproductive (sexual) period to the next.
Climax community	The community developed in a stable habitat. A community in equilibrium.
Community	A collection of populations of animals and plants that occur naturally together in a common environment.
Cyanobacteria	Blue-green algae.
Ecosystem	The whole interacting organismic and physico-chemical environment in which the species live.
Epilithon	The community living on stone (and other solid) surfaces in water.
Epipelon	The community living on soft, penetratable (silt/mud) deposits.
Epiphyton	The community living on underwater plant surfaces.
Epipsammon	The community living attached to sand grains.
Euryoecious	Having a tolerance to a wide range of conditions.
Eurysaprobic	Referring to distribution over a wide range of organic matter content.
Eutrophication	The increase in supply of nutrients to an environment.
Frustule	The term for the whole of the encasing siliceous 'plates' which surround the diatom protoplast.
Girdle band(s)	The 'hoops' which connect the two diatom valves together to make a complete cover for the protoplast.
Habitat (niche)	The specific place where species or communities live.
Indicator species	Species (kinds) of organisms (eg diatoms) which, when found, indicate some special physico-chemical feature of the environment.
Ion exchange	The exchange of chemicals (ions) between organisms and the environment.
Macroinvertebrate	Animals more complex than protozoa, but less complex than the skeletal (internal) groups. Worms, insects, snails, etc.
Macrophyte	Plants that are easily visible with the naked eye.
Mesosaprobic	Waters of intermediate organic matter status.
Metaphyton	The loose community of organisms associated with surfaces such as plant stems.
Mucilage	A complex glutinous carbohydrate secreted by certain plants, for example diatoms.
Oligosaprobic	Waters with only a small amount of organic matter.
Pedunculate	Possessing a stalklike structure as a means of attachment to the surface on which the organism lives.
Pigment diversity	Diatoms contain chlorophyll but also a whole range of brown pigments (hence diverse) which overwhelm the green of the chlorophyll.
Potamon	The slow-flowing region of a river.
Polysaprobic	Waters containing excessive amount of organic material.
Primary valves	The first valves formed within the germinating auxospore. These are of a different form (or distorted form) compared with the succeeding valves produced at each cell division of a diatom.
Punctum (puncta)	Small hole (appearing as a dot under the microscope).

Phithron	The fast-flowing region of a river.
Riffle	A region (shallow) in a river where the water flows over pebbles, boulders, cobbles.
Saprobity	Classification of a species depending on organic enrichment according to the Continental Saprobic System.
Saprophilous	Occurring in waters of moderate organic matter content.
Saproxenous	Occurring in waters of extremely low organic matter content.
Sedgwick Rafter cell	A covered counting cell, for use under the microscope, which is marked with a grid and holds a standard volume of liquid.
Semi-quantitative	Data based on simple counting techniques which aim to give a broad picture.
Stigma	An isolated pore in the central area of a diatom valve.
Striae	The striped markings, as seen under the microscope, on diatom valves.
Taxon (Taxa)	A unit of classification of any rank in the hierarchial scale.
Ultrasonic	High frequency sound (used to detach diatoms from surfaces or break open diatom frustules).
Unicellular	Organisms which exist as single cells.
Valve	The upper or lower encasing element of a diatom frustule.
Xenosaprobic	Waters containing almost no organic matter.

8 References

Aloi, J E, Loeb, S L & Goldman, C C (1988). Temporal and spatial variability of the eulittoral epilithic periphyton, Lake Tahoe, California-Nevada. *J. Freshwat. Ecol.* **4**, 401–40

Antoine, S E & Benson-Evans, K (1985). Colonisation rates of benthic algae on four different rock substrata in the River Ithon, Mid Wales, UK *Limnologica* **16**, 307–313

Archibald, R E M (1972). Diversity in some South African diatom associations and its relation to water quality. *Wat. Res.* **6**, 1229–1238

Archibald, R E M (1983). The diatoms from the Sundays and Great Fish Rivers in the Eastern Cape Province of S Africa. *Bibliotheca Diatomologica* **1**, 432 pp.

Backhaus, D (1968). Ökologische Untersuchungen an den Aufwuchsalgen der obersten Donau und ihrer Quellflüsse. II–IV. *Arch. Hydrobiol.* Suppl. **34**, 24–73, 130–149, 251–320

Battarbee, R (1973). A new method for the estimation of absolute microfossil members with special reference to diatoms. *Limnol. & Oceanogr.* **18**, 647–653

Behre, K (1966). Zur Algensoziologie des Süsswassers. *Arch. Hydrobiol.* **62**, 125–164

Besch, W K, Ricard, M & Cantin, L (1972). Benthic diatoms as indicators of Mining Pollution in the northwest Mirimichi River System, New Brunswick, Canada. *Int. Rev. ges. Hydrobiol.* **57**, 39–74

Biggs, B J F (1988). Artificial substrate exposure times for periphyton biomass estimates in rivers. *New Zealand Journal of Marine and Freshwater Research* **22**, 507–575

Blinn, D W, Frederickson, A & Korte, V (1980). Colonisation rates and community structure of diatoms on three different rock substrata in a lotic system. *Br. phycol. J.* **15**, 303–310

Blum, J L (1956). The ecology of river algae. *Bot. Rev.* **22**, 291–341

Bursik, H (1973). Aufwuchsbiozönosen auf Pontans in der Donau bei Wien. Doctoral Diss. Universität Wien.

Butcher, R W (1932). Studies in the ecology of rivers II. The microflora of rivers with special reference to the algae on the river bed. *Ann. Bot.* **46**, 813–861

Butcher, R W (1940). Studies in the ecology of rivers IV. Observations on the growth and distribution of the sessile algae in the River Hull, Yorkshire. *J. Ecol.* **28**, 210–233

Butcher, R W (1946). Studies in the ecology of rivers VI. Algal growth in certain highly calcareous streams. *J. Ecol.* **33**, 268–283

Butcher, R W (1947). Studies on the ecology of rivers VIII. The algae of organically enriched waters. *J. Ecol.* **35**, 186–191

Cairns, J, Albough, D W, Busey, F & Chanay, M D (1968). The Sequestrial Comparison on Index—a simplified method for non-biologists to estimate relative differences in biological diversity in stream pollution studies. *J. Wat. Pollut. Control Fed.* **40**, 1607–13

Cambra, J (1987). Flore et végétation algologiques des eaux épicontinentals de la Resérve Naturelle des vallées d'Ordesa (Pyrénées aragonaises). *Candollea* **42**, 475–490

Casey, H, Clarke, R T & Marker, A F H (1981). The seasonal variation in silicon concentratrion in chalk-streams in relation to diatom growth. *Freshwater Biology* **11**, 335–344

Caspers, H & Schultz, H (1960). Studien zur Wertung der Saprobiensysteme. *Int. Rev. ges. Hydrobiol.* **45**, 535–565

Caspers, H & Schultz, H (1962). Weitere Unlagen zur Prüfung der Saprobiensysteme. *Int. Rev. ges. Hydrobiol.* **47**, 100–117

Cattaneo, A, Ghittori, S & Vendegna, V (1975). The development of benthonic photocoenosis on artificial substrates in the Ticino River. *Oecologia* **19**, 315–327

Cazaubon, A (1988). Rôle du courant sur la microdistribution des diatomées épilithiques dan une riviére Méditerranéene, L'Argens (Var. Provence). *Proc. 9th Diatom Symposium (1980)* (ed. F E Round), 93–107. Biopress (Bristol) & Koeltz, Koenigstein

Cazaubon, A & Loudiki, M (1983). Microrépartition des algues épilithiques sur les cailloux d'un torrent Corse, Le R Rizzanère. *Annls. Limnol.* **22**, 3–16

Cook, S E K (1976). Quest for an index of community structure sensitive to water pollution. *Environ. Pollut.* **11**, 26–9

Coste, M (1976). Contribution à l'écologie des diatomées benthiques et périphytiques de la Seine: Distribution longitudinale et influence des pollutions. *Soc. Hydrotec. de France, Journ. de l'Hydraulique* **14**, Rapp. 9, 1–7

Coste, M (1984). Opération Seine Rivière propre. Evaluation de la qualité hydrobiologique: Pouissons—Diatomées. Rapp. Agon. Seine Normandie, Corseil Regional Ile de France (CEMAGREF), 35 pp.

Coste, M (1990) Diagnostic biologique de la qualité des eaux continentales: les principales méthodes microfloristiques. DEA Hydroécologie Hydrobiologie de la qualité de l'eau. Cemagref Bordeaux.

Cushing, C E, McIntyre, C D, Cummins, K W, Minshall, G W, Peterse, N R C, Sedell, J R & Van Note, R L (1983). Relationships among chemical, physical and biological indices along river continua based on multivariate analyses. *Arch. Hydrobiol.* **98**, 317–376

Dell'Uomo, A (1981). Alghe di acque correnti e loro impiego come bioindicateri. *Giorn. Bot. Ital.* **115**, 327–342

Denisger, J, Austin, A & Lucey, W P (1986). Periphyton communities in a pristine mountain stream above and below heavy metal mining operations. *Freshwater Biology* **16**, 209–218

De Nicola, D M, McIntire, C D, Lamberti, G A, Gregory, S V & Ashkenas, L R (1990). Temporal patterns of grazer-periphyton interactions in laboratory streams. *Freshwater Biology* **23**, 475–489

Descy, J P (1976). La végétation algale benthique de la Somme (France) et ses relations avec la qualité des eaux. *Mém.* Soc. Roy. Bot. Belg. **7**, 101–128

Descy, J P (1979). A new approach to water quality estimation using diatoms. *Nova Hedwigia, Beiheft* **64**, 305–323

Descy, J P (1980). Utilisation des algues benthiques comme indicateurs biologues comme indicateurs biologiques de la qualité des eaux courantes. In: *La pollution des eaux continentales. Incidence sur les biocénmoses aquatiques* (ed. P Pesson). Gauthier-Villars, Paris. pp. 169–194

Descy, J P (1983). Les communautés algales de la haute Meuse Belge. et la qualité de ses eaux: Un bilan à l'automne 1981. *Bull. Soc. Roy. Bot. Belg.* **116**, 31–50

Descy, J P (1984). Écologie et distribution de Diatomées benthiques dans le Bassin Belge de la Meuse. *Doc. de Travail.* **18**. *Inst. Roy. Sci. Nat. Belge.* 25 pp. 70 pl.

Descy, J P & Coste, M (1987). Utilisation des diatomées benthique pour l'évolution de la qualité des eaux courant. *Premiere Rapp. d'avancement a UNCED*, 24 pp.

Descy, J P & Coste, M (1988). Utilisation des diatomées benthique pour l'évolution de la qualité des eaux courant. *2eme Premiere Rapp. d'avancement a UNCED*, 49 pp.

Descy, J P & Coste, M (1990). Utilisation des diatomées benthiques pour l'évoluation de la qualité des eaux courantes. *Rapport Final Contract CEE B–71–23*, 64 pp.

Descy, J-P, Empain, A Lambinon, J (1982). Un inventaire de la qualité des eaux du bassin wallon de la Meuse (1976–1980). *Trib. Cebedeau. Nos* 463–464, **35**, 267–278

Douglas, B (1958). The ecology of the attached diatoms and other algae in a small stony stream. *J. Ecol.* **46**, 295–322

Duffer, W R & Dorris, T C (1966). Primary productivity in a southern great plains stream. *Limnol. Oceanogr.* **11**, 143–157

Dussart, G. (1982). A pragmatic approach to water quality microbiology in a clean river area. *Environ. Poll. (Ser. A)* **27**, 97–118

Economou-Amilli, A (1980). Periphyton analysis for the evaluation of water quality in running waters of Greece. *Hydrobiologia* **74**, 39–48

Edwards, R W, Benson-evans, K, Learne, M A, William, P & Williams, R (1972). A biological survey of the River Taff. *Wat. Pollut. Control* **71**, 144–166

Eggleston, D M, Wilson, R T & Harrington, C B (1987). The effect of two buffers on cell yield of *Nitzschia palea* (Kütz.) W Smith. *J. Freshwat. Ecol.* **4**, 33–35

Eichenberger, E & Wuhrmann, K (1975). Growth and photosynthesis during the formation of a benthic algal community. *Verh. Int. Verein. Limnol.* **19**, 2035–2042

Engelberg, K (1987). Die Diatomeen-Zönose in einem Mittelgebirgsbach und die Abgrenzung jahreszeitlicher Aspekte mit Hilfe der Dominanz-Identität. *Arch. Hydrobiol.* **110**(2), 217–236

Evans, G H & Marcan, E (1976). Diatom community response to variations in effluent concentration. *Environ. Poll.* **10**, 115–126

Evenson, W E, Rushforth, S R, Brotherson, J D & Fungladda, N (1981). The effects of selected physical and chemical factors on attached diatoms in the Uintah Basin of Utah, USA. *Hydrobiologia* **83**, 325–330

Fabri, R & Leclercq, L (1979). Flore et végétation algales des ruisseaux de la Réserve naturelle domaniale des Hautes Fagnes (province de Liège, Belgique). *Min de l'Agric. Admin. des Eaux et Forêts. Trav. no.* **11**, 47 pp

Fairchild, G W & Lowe, R L (1984). Artificial substrates which release nutrients: effects on periphyton and invertebrate succession. *Hydrobiologia* **114**, 29–37

Fairchild, G W, Lowe, R W & Richardson, W B (1985). Algal periphyton growth on nutrient-diffusing substrates: an *in situ* bioassay. *Ecology* **66**, 465–472

Fisher, S G, Gray, L J, Grimm, N B & Busch, D B (1982). Temporal succession in a desert stream ecosystem. *Ecol. Monogr.* **52**, 93–110

Fjerdingstad, E (1964). Pollution of streams estimated by benthal phytomicroorganisms. 1. A saprobic system based on communities of organisms and ecological factors. *Int. Rev. ges. Hydrobiol.* **49**, 63–131

Fjerdingstad, E (1965). Taxonomy and saprobic valency of benthic phytomicroorganisms. *Int. Rev. ges. Hydrobiol.* **50**, 475–604

Fritsch, F E (1948) The structure and reproduction of the algae. Vol 1 Cambridge University Press.

Fryer, G (1987) Quantitative and qualitative: numbers and reality in the study of living organisms. Freshwater Biology **17** 177–189

Furse, M T, Moss, D, Wright, J F & Armitage, P D (1984). The influence of seasonal and taxonomic factors on the ordination and classification of running-water sites in Great Britain and on the prediction of their macro-invertebrate communities. *Freshwat. Biol.* **14**, 257–280

Gale, W F (1975). Ultrasonic removal of epilithic algae in a bar-clamp sampler. *J. Phycol.* **11**, 472–473

Gale, W F, Gurzynski, A J & Lowe, R L (1979). Colonization and standing crops of epilithic algae in the Susquehanna River, Pennysylvania. *J. Phycol.* **15**, 117–123

Goldsborough, L G (1989). Examination of two dimensional spatial patterns of periphytic diatoms using an adhesive surficial peel technique. *J. Phycol.* **25**, 133–143

Gumtow, R B (1955). An investigation of the periphyton in a riffle of the West Gallatin River, Montana. *Trans Am. Microsc. Soc.* **74**, 278–292

Hartmann, D (1983). Beziehungen zwischen der Diatomeen-Flora und dem Wasserchemismis in Fliessgewässern des Sauerlandes. 1. Die Ruhr. *Ber. naturhist. Ges. Hannover* **126**, 91–135

Hawkes, H A (1982). Biological surveillance of rivers. *Wat. Pollut. Control* **81**, 329–342

Hellawell, J M (1972). Change in natural and managed ecosystems: detection, measurement and assessment. *Proc. Roy. Soc.* **B 197**, 31–56

Hill, W R & Knight, A W (1987). Experimental analysis of the grazing interaction between a mayfly and stream algae. *Ecology* **68**, 1955–1965

Hoagland, K D, Roemer, S C & Rosowski, J R (1982). Colonization and community structure of two periphyton assemblages with emphasis on the diatoms (Bacillariophyta). *Am. J. Bot.* **69**, 188–213

Hodgkiss, I J & Kan, H W P (1978) A comparison of four biological indices of pollution based on studies of streams in the New Territories. Hong Kong, Water Pollution Control **77** 532–538

Hodgkiss, I J & Law, C Y (1985). Relating diatom community structure and stream water quality using species diversity indices. *Wat. Pollut. Control* **84**, 134–139

Hornung, H (1959). Floristisch-ökologische Untersuchungen an der Echaz. unter besonderer Berücksichtigung der Vereunreinigung durch Abwässer. *Arch. Hydrobiol.* **55**, 52–126

House, M E (1989) A water quality index for river management. J I W F M **3** 336–344

Hudson, C & Bourget, C (1983) The effect of light on the vertical structure of epibenthic diatom communities. *Bot. Mar.* **26**, 317–330

Hulbert, S H (1971) The non-concept of species diversity: A critique and alternative parameters. Ecology **52** 577–586

Hürlimann, J & Straub, F (1991). Morphologische und ökologische characterisierung von Sippen um den *Fragilaria capucina*-komplex sensu Lange-Bertalot 1980. *Diatom Research*, **6**, 21–47

Hustedt, F (1927–66) Die Kieselalgen. In: *Rabenhorst's Kryptogamenflora von Deutschland, Österreich und der Schweiz* Leipzig. Academische Verlagsgesellschaft.

Hustedt, F (1930). Bacillariophyta (Diatomeae). In: *Die Süsserwasserflora Mitteleuropa* (Pascher, A., ed.), **10**, 466 pp. 2nd ed. Fischer, Jena.

Iserentant, R Blanke, D (1986). Sensibilité des peuplements de diatomeées aux changements de qualité d l'eau. Une essai le transplantation en eau courante. In: *Proc. 8th Int. Diatom Symposium, Paris* (ed. M Ricard), 347–354.

Jahn, R & Wendker, S (1987). Untersuchungen zur Diatomeenflora der Werra—einem extrem salzbelasteten Biotop. *Nova Hedw.* **44**, 163–173

Kadlubowska, J Z (1985). Untersuchungen der Stetigkeit der Diatomeen gesellschaft aus der Salzquelle, den Limnokrenen und dem Moortempel. *Verh. Int. Verein. Limnol.* **22**, 2834–2837

Kaesler, R L, Cairns, J Jr & Crossman, J S (1974). Redundancy in data from stream surveys. *Wat. Res.* **8**, 637–642

Katoh, K (1991). A comparative study on some pollution indices. *Diatom* **6**, 11–17

Kawecka, B (1981). The ecological characteristics of diatom communities in the mountain streams of Europe. *Proc. 6th Diatom Symposium on Recent and Fossil Diatoms* (Ed. R Ross), 425–433

Keithan, E D, Lowe, R L & DeYoe, H R (1988) Benthic diatom distribution in a Pennsylvania stream: Role of pH and nutrients. J Phycol **24** 581–585

Kobayasi, H & Mayama, S (1982). Most pollution-tolerant diatoms of severely polluted rivers in the vicinity of Tokyo. *Jap. J. Phycol.* **30**, 188–196

Kolkwitz, R & Marsson, M (1908). Ökologie der pflanzlichen Saprobien. *Ber. Deutsche Bot. Ges.* **26a**, 505–519

Kothé, P (1962). Der Ärtenfehlbetrag, ein einfaches Gütekriterium und seine Anwendung bei biologischer Vorflutenuntersuchungen. *Dt. Gewässerkundl. Mitt* 6, 60–65

Krammer, K & Lange-Bertalot, H (1985). Naviculaceae, Neue und wenig bekannte Taxa, neue kombinationen und Synononyme sowie Bemerkungen zu einigen Gattingen. *Bibliotheca Diatomologica* **9**, 230 pp.

Krammer, K & Lange-Bertalot, H (1986). *Süsswasserflora von Mitteleuropa. Bacillariophyceae 2/1 Naviculaceae.* 876 pp. Gustav Fischer, Stuttgart.

Krammer, K. & Lange-Bertalot, H. (1987). Morphology and taxonomy of *Surirella ovalis* and related taxa. *Diatom Research* **2**, 77–8

Krammer, K. & Lange-Bertalot, H. (1988). *Süsswasserflora von Mitteleuropa. Bacillariophyceae 2/2 Epithemiaceae, Surirellaceae.* 596 pp. Gustav Fischer, Stuttgart

Krammer, K & Lange-Bertalot, H (1991a). *Süsswasserflora von Mitteleuropa. Bacillariophyceae 2/3 Centrales, Fragilariaceae, Eunotiaceae.* 576 pp. Gustav Fischer, Stuttgart

Krammer, K & Lange-Bertalot, H (1991b) 2/4 Achnanthaceae, Kritische Erganzungen zu *Navicula* (Lineolatae) und *Gomphonema*. Gesamt-literaturverzeichnis Teil 1–4 Gustav Fischer, Stuttgart.

Lamberti, G A & Resh, V H (1985). Comparability of introduced tiles and natural substrates for sampling lotic bacteria, algae and macroinvertebrates. *Freshwat. Biol.* **15**, 21–30

Lamkin, A J III & Sommerfeld, M R (1982). Algal distribution in a small, intermittent stream receiving acid mine drainage. *J. Phycol.* **18**, 196–199

Lange-Bertalot, H (1979). Pollution tolerance as a criterion for water quality estimation. *Nova Hedwigia, Beiheft* **64**, 285–304

Lange-Bertalot, H & Ruppel, M (1980). Zur Revision taxonomische problematischer, ökologisch jedoch wichtiger Sippen der Gattung *Achnanthes* Bory. *Arch. Hydrobiol. Suppl.* **60**, *Algol. Stud.*, **26**, 1–31

Lay, J A & Ward, A K (1987). Algal community dynamics in two streams associated with different geological regions in the south eastern United States. *Arch. Hydrobiol.* **108**, 305–324

Leclercq, L (1977). Végétation et caractéristiques physico-chimiques de deux rivières de haute Ardenne (Belgique): La Helle et la Roer supérieure. *Lejeunia* NS **88**, 1–42

Leclercq, L (1988). Utilisation de trois indices, chimique, diatomique et biocénotique pour l'évaluation de la qualité de l'eau de la Joncquière, rivière calcaire polluée par le village de Douche (Belgique, Prov. Namur). *Mém. Soc. Roy. Bot. Belg.* **10**, 26–34

Leclercq, L & Manquet, B (1987). Deux nouveaux indices chimique et diatomique de qualité d'eau courante. Application au Samson et à ses affluents. Comparaison avec d'autres indices chimiques, biocenotiques et diatomiques. *Inst. Roy. Sc. Natur. Belg. Documents de travail* **38**, 113 pp.

Lee, K & Round, F E (1989) Studies on freshwater *Amphora* species. III. *Amphora pediculus* (Kütz.) Grun. and some possibly related forms. *Diatom Research* **4**, 79–87

Leland, H V & Carter, J L (1984). Effects of copper on species composition of periphyton in a Sierra Nevada, California Stream. *Freshwat. Biol.* **14**, 281–296

Levandowsky, M (1972). A ordination of phytoplankton populations in ponds of varying salinity and temperature. *Ecology* **53**, 398–407

Lock, M A, Wallace, R R, Costerton, J W, Ventullo, R M & Charlton, S E (1984). River epilithon: towards a structural functional model. *Oikos* **42**, 10–22

Lowe, R L (1974). Environmental requirements and pollution tolerance of freshwater diatoms. *USEPA Environmental monitoring series 640/4-74-065. Cincinnati National environ. Res. Center*, 334 pp.

Margalef, R (1949). A new limnological method for the investigation of thin-layered epilithic communities. *Hydrobiologia* **1**, 215–216

Margalef, R (1958). 'Trophic' typology versus biotic typology as exemplified in the regional limnology of northern Spain. *Ver. Inter. Ver. Limnologie* **13**, 339–49.

Marker, A F H (1976). The benthic algae of some streams in southern England. I. Biomass of the epilithon in some small streams. *J. Ecol.* **64**, 343–358

Mayama, S & Kobayasi, H (1984). The separated distribution of the two varieties of *Achnanthes minutissima* Kuetz. according to the degree of river water pollution. *Jap. J. Limnol,* **45**, 304–312

McIntire, C D (1968). Structural characteristics of benthic algal communities in laboratory streams. *Ecology* **49**, 520–537

Moller, B & Pankow, H (1981). Algensoziologische und saprobiologische untersuchung an Varflutern der Elbe. *Limnologica* (Berlin) **13**, 291–350

Murphy, P M (1978). The temporal variability in biotic indices. *Environ. Poll.* **17**, 227–236

Negoro, K I & Gotoh, T (1983). Diatom vegetation of the River Yura. *Mem. Fac. Agr. Kinki Univ.* **16**, 67–118

Oemke, M P & Burton T M (1986). Diatom colonization dynamics in a lotic system. *Hydrobiologia* **139**, 153–166

Olive, J H & Price, J L (1977). Diatom assemblages of Cuyahoga River, NF, Ohio (USA). *Hydrobiologia* **57**, 175–187

Palmer, C M (1969). A composite rating of algae tolerating organic pollution. *J. Phycol.* **5**, 78–82

Pantle, R & Buck, H (1952). Die biologische Überwachung der Gewässer and die Darstelling der Ergebnisse. *Gas und Wasserfach.* **96**, 604

Patrick, R (1949). A proposed biological measure of stream conditions, based on a survey of the Conestogon basin, Lancaster Country, Pennsylvania. *Proc. Acad. Nat. Sci. Philad.* **101**, 272–341

Patrick, R Reimer, C W (1966). The diatoms of the United States exclusive of Alaska & Hawaii. Vol. 1. *Monogr. Acad. Nat. Sci., Philadelphia.* No. **13**, 688 pp.

Patrick, R & Reimer, C W (1975). The diatoms of the United States exclusive of Alaska & Hawaii. Vol. 2. *Monogr. Acad. Nat. Sci., Philadelphia.* No. **13**, 213 pp.

Patrick, R & Roberts, N A (1979). Diatom communities in the middle Atlantic States, USA. Some factors that are important to their structure. *Proc. 5th Symp. on Recent and Fossil Diatoms. Nova Hedw. Beih.* **64**, 265–283

Peterson, C G & Stevenson, R J (1989). Substratum conditioning and diatom colonization in different current regimes. *J. Phycol.* **25**, 790–793

Pringle, C M & Bowers, J A (1984). An *in situ* substratum fertilization technique: diatom colonization nutrient-enriched on sand substrata. *Can. J. Fish. Aquat. Sci.* **41**, 1247–1251

Raabe, H (1951). Die Diatomeenflora der ostholsteinisalen Fliessgewässer. *Arch. Hydrobiol.* **44**, 521–638

Ramm, A E (1988). The community degradation index: A new method for assessing the deterioration of aquatic habitats. *Wat. Res.* **22**, 293–300

Reichardt, E (1984). Die Diatomeen der Altmühl. *Bibliotheca Diatomologica* **6**, 169 pp.

Reisen, W K & Spencer, D J (1970). Succession and current demand relationships of diatoms on artificial substrates in Praters Creek, South Carolina. *J. Phycol*, **6**, 117–121

Reiter, M A (1986). Interactions between the hydrodynamics of flowing water and the development of a benthic algal community. *J. Freshwat. Ecol.* **3**, 511–517

Reynolds, C S (1988). Potamoplankton: Paradigms, Paradoxes and Prognoses. In: *Algae and the aquatic environment* (Ed F E Round), pp 285-311. Biopress Ltd Bristol, UK

Robinson, C T & Rushforth, S R (1987). Effects of physical disturbance and canopy cover on attached diatom community structures. *Hydrobiologia* **154**, 49–59

Round, F E (1989). Diatoms as water quality indicators in British rivers. *DOE report, Contract no. 120021-00-37.*

Round, F E (1990). The effect of liming on the benthic diatom population in three upland Welsh lakes. *Diatom Research* **5**, 129–140

Round, F E (1991). Epilithic diatoms in acid streams flowing into the reservoir Llyn Brianne. *Diatom Research* **6**, 137–145

Round, F E, Crawford, R M & Mann, D G (1990). The diatoms. Biology and morphology of the genera. Cambridge University Press, 747 pp.
Rumeau, A & Coste, M (1988). Initiation a la sysématique des Diatomées d'eau douce. Pour l'utilisation pratique d'un indice diatomique génerique. *Bull. Fran. de la pêche et de la pisciculture*, no. **109**, 69 pp.
Rushforth, S R, Squires, L E & Cushing, C E (1986). Algal communities of springs and streams in the Mt St Helens region, Washington, USA following the May 1980 eruption. *J. Phycol.* **22**, 129–137
Sabater, S & Sabater, F (1988). Diatom assemblages in the River Ter. *Arch. Hydrobiol.* **111**, 397–408
Sabater, S, Sabater, F & Tomas, X (1987). Water quality and diatom communities in two Catalan rivers (NE Spain). *Water Res.* **21**, 901–911
Salden, N (1978). Beiträge zur Ökologie der Diatomeen des Süsswassers. *Decheniana* **22**, 1–238
Scharf, W (1984). Kieselalgen als Indikatoren für die Beurteilung der Ablaufqualität von Kläranlagen. *Forum Städe-Hygiene. Zeitschrift für Forschung und Technik in der Wasser-, Boden-und Luft-Hygiene* **35**, 268–273
Schoeman, F R (1976). Diatom indicator groups in the assessment of water quality in the Juskei-Crocodile river system (Transvaal, Republic of South Africa). *J. Limnol. Soc. Southern Africa* **2**, 21–24
Sládeček, V (1973). System of water quality from the biological point of view. *Ergebn. Limnol.* **7**, IV, 1–218
Sládeckova, A (1962) Limnological investigation methods for the periphyton (Aufwuch's) community. *Bot. Rev.* **28**, 286–350
Statzer, B & Higler B (1986). Stream hydraulics as a major determinant of benthic invertebrate zonation patterns. *Freshwat. Biol.* **16**, 127–139
Steinberg, C & Schiefele, S (1988). Biological indication of trophy and pollution of running waters. *Zeitschrift für Wasser- Abwasser-Forschung*, **21**, 227–234
Steinman, A D & Lamberti, G A (1988). Lotic algal communities in the Mt St Helens region six years following the eruption. *J. Phycol.* **24**, 482–489
Steinman, A D & McIntire, C D (1987). Effects of irradiance on the community structure and biomass of algal assemblages in laboratory streams. *Can. J. Fish. Aquat. Sci.* **44**, 1640–1648
Stevens, S S (1946). On the theory of scales of measurement. *Science* NY **103**, 622–680
Stevenson, R J (1983). Effects of current and conditions stimulating autogenically changing microhabitats on benthic diatom immigration. *Ecology* **64**, 1514–1524
Stevenson, R J (1984a). How currents on different sides of substrates in streams affect mechanisms of benthic algal accumulation. *Int. Rev. Ges. Hydrobiol.* **69**, 241–262
Stevenson, R J (1984b). Epilithic and epipelic diatoms in the Sandusky River, with emphasis on species diversity and water pollution. *Hydrobiologia* **114**, 161–175
Stevenson, R J & Hashim, R J (1989). Variation in diatom community structure among habitats in sandy streams. *J. Phycol.* **25**, 678–686
Sumita, M & Watanabe, T (1983). New general estimation of river pollution using new diatom community index (NDCI) as biological indicators based on specific composition of epilithic diatom communities—Applied to Asano-gawa and Sai-gawa rivers in Ishikawa prefecture. *Jap. J. Limnol.* **44**, 329–340
Symoens, J J (1957). Les eaux douces de l'Ardenne et des régions voisines: les milieux et leur végétation algale. *Bull. Soc. roy. Bot. Belg.* **89**, 111–34
Thorpe, M W & Williams, I L (1980). A guide to the use of algae for the biological surveillance of rivers. *DOE, Water Data Unit, Technical Memorandum* **20**, 34 pp.
Verneaux (1981). Fondements biologiques et écologiques de l'étude de la qualité des eaux continentàles. Principales méthodes biologiques. In: *La pollution des eaux continentales. Incidences sur les biocenoses aquatiques* (Pesson, P, ed), 229–285
Washington, N G (1984). Diversity, biotic and similarity indices. A review with special relevance to aquatic ecosystems. *Wat. Res.* **18**, 653–694
Watanabe, (1986). Saprophilous and eurysaprobic diatom taxa to organic water pollution and diatom assemblage index (DAIpo). *Diatom* **2**, 23–73
Watanabe, T, Asai, K & Houki A (1988). Numerical index of water quality using diatom assemblages. In: *'Biological monitoring of Environmental Pollution'* (eds M Yasuno & B A Whitton), pp 179–192. Tokai University Press.
Wegl, R (1983). Index für die Limnosaprobität (1983). *Wasser und Abwasser* **26**, 175 pp.

Whitton, B A & Diaz, B M (1981). Influence of environmental factors on photosynthetic species composition in highly acidic waters. *Verh. Internt. Verein. Limnol.* **21**, 1459–1465

Whitton, B A, Rott, E & Friedrich, G (ed) (1991) Use of algae for monitoring rivers. Dr E Rott, Inst für Botanik, Universität Innsbruck, Osterreich.

Youngman, R E (1971). Algal monitoring of water supply reservoirs and rivers. *Water Research Centre, Technical Memorandum TM 63*. WRC.

Zelinka, M & Marvan, P (1961). Zur Präsisierung der biologischen Klassification der Reinheit fliessender Gewässer. *Arch. Hydrobiol.* **57**, 389–407

Address for correspondence

Correspondence about this booklet or any method in this series should be addressed to:

The Secretary
Standing Committee of Analysts
Department of the Environment
Drinking Water Inspectorate
Romney House
43 Marsham Street
London
SW1P 3PY

Correspondence of a technical nature concerning this booklet should be addressed to:

Professor F E Round
Department of Botany
University of Bristol
Bristol
BS8 1UG

Department of the Environment

Standing Committee of Analysts

Members of the committee assisting with this booklet

Adam, J D	3
Basketter, F B	3
Bellinger, E G	2
Benson-Evans, K	1
Best, G A	3
Bolas, P M	2
Croft, B T	3
Davidson, I O M	3
Descy, J P	2
Finniear, M C	3
Harding, J P C	1
Haslam, J M	2
Hawkes, H A	2
Hellawill, J M	2
Howard, S	1
Hurcombe, M R	3
Jefferey, J	3
Jones, J G	3
Martin, R	2
Marker, A F	1
McDonnell, E	3
Mycock, E R	2
Pittwell, L R	1,3
Purdie, L D	3
Riley, J P	3
Round, F E	1*
Sweeting, R	1
Taylor, D	3
Thompson, K C	3
Toms, I	2
Vincent, A J	3
Ware, A	3
Westlake, D F	2
Whitton, B A	1
Williams, D A	3
Wheatstone, K	3

1. Special WG7 Panel and WG7
2. Active in WG7
3. Active in Main Committee

* Author

Printed in the United Kingdom for HMSO
Dd296009 6/93 C7 G531 10170